# MACBOOK PRO M4 USER GUIDE

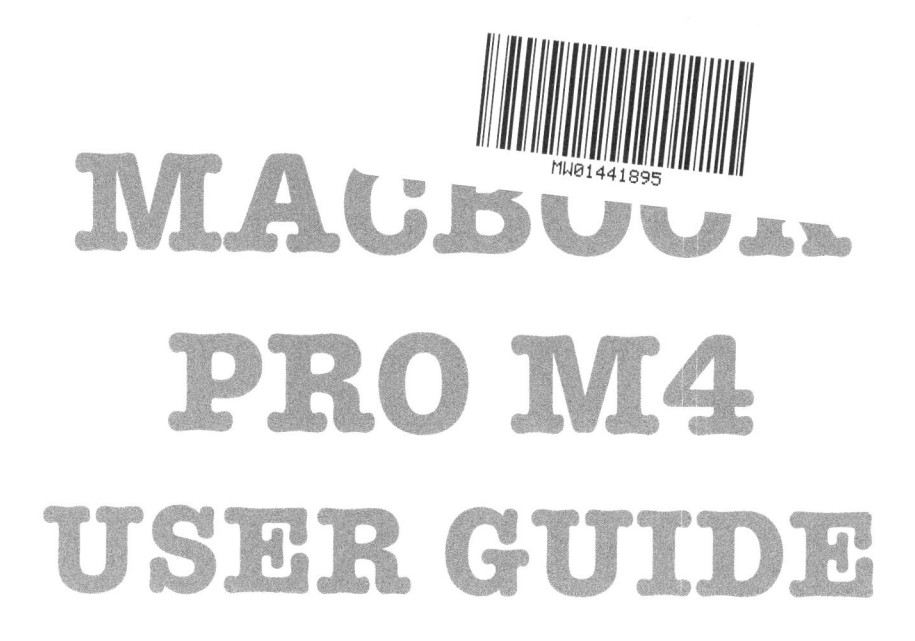

Beginner Seniors Manual to Use and Master the 14" & 16" Mac Laptop (2024 release) with M4 chip plus Tips & Tricks on macOS Sequoia and Apple Intelligence

**Leandro Barnes**

**Copyright © 2024 by Leandro Barnes**

All rights reserved. Apart from brief quotations in critical reviews or other non-commercial uses, no part of this book may be published, reproduced including: any electronic, photostating, or mechanical means, without the prior written consent of the author or publishers, unless it's otherwise lawfully permitted by the author's express permission.

Country where the product is printed: United States of America

For further inquiries or requests including permission and seizure please contact the Author or Publisher.

# TABLE OF CONTENTS

TABLE OF CONTENTS .................................. 3

CHAPTER 1: APPLE MACBOOK PRO GETS SPEED BOOST AND ANTI-GLARE SCREEN ...................................................... 16

PRICE AND CONFIGURATION OPTIONS OF MACBOOK PRO 2024 ........................................... 16

NEW CHIPS IN TOWN: IMPROVED DISPLAY AND CAMERA FEATURES ................................... 17

Advanced connectivity options and ways of setting up your MacBook Pro 2024 ................................ 19

COMPARING THE M4 PRO AND M4 MAX: PERFORMANCE AND TECHNOLOGY ADVANCEMENTS ............................................. 21

CHAPTER 2: SETTING UP YOUR NEW MACBOOK .................................................. 25

Set Location and Accessibility Settings ............... 25

Connect to Wi-Fi ................................................ 26

Transfer Data with Migration Assistant ............. 27

Set Up Apple ID and Login .................................. 27

Create a Computer Account ................................ 28

Know the Find My App ....................................... 29

Make This Your New Mac ................................... 29

3

Enable and Configure Siri ................................. 31
Set Up FileVault for Disk Encryption ................ 33
Set Up Touch ID (If Available) ........................... 34
Choose a Desktop Theme .................................. 34
Update Your Mac................................................35

## CHAPTER 3: NAVIGATING THE MAC OS INTERFACE ................................................ 38

Understanding the Layout: The Desktop and the Menu Bar .......................................................... 38
   Desktop Functionality ..................................... 39
   The Menu Bar .................................................. 39
   The Apple Menu............................................... 40
   App Menu..........................................................41
   The Help Menu ................................................ 42

## CHAPTER 4: THE FINDER ........................ 44

Opening a Finder Window ................................. 44
Quick View and Editing Files ............................. 44
Organizing .......................................................... 45
Syncing Devices ................................................. 45
The Finder Sidebar ............................................ 46
Adjusting File and Folder Views ........................47
Using the Preview Pane .................................... 49
Gallery View........................................................ 50

Quick Actions..................................................... 50
Quick Look.......................................................... 51
Accessibility Tip..................................................52
Get There Faster With The Go Menu..................52

## CHAPTER 5: THE DOCK ON YOUR MAC 54

How to Open an Application or Open a File...... 54
Closing an App....................................................55
Adding an Item to the Dock .............................. 56
Removing an Item from the Dock.......................57
Mission Control: See Everything Open on Your Mac .....................................................................57
Viewing All Open Windows in an App ............... 58
Customizing the Dock ....................................... 59
Additional Tips for Using the Dock.................... 59

## CHAPTER 6: NOTIFICATION CENTER ON YOUR MAC................................................... 61

Accessing Notification Center ............................61
Interacting with Notifications ............................ 63
Customize Notification Preferences................... 64
Setting Your Notification Settings ..................... 65
    Launch notification settings:......................... 65
    Setting Up Notifications ............................... 66

## CHAPTER 7: FOCUS .................................... 68

5

Setting Up Focus.................................................... 68
Enabling and Disabling Focus............................ 69

## CHAPTER 8: ACCESSING AND CUSTOMIZING YOUR WIDGETS ............. 71

Step 1: Opening the Widget Gallery ................ 71
Adding Widgets................................................. 71
Step 3: Removing Widgets.................................72
Step 4: Rearranging Widgets.............................73
ADDING THIRD-PARTY WIDGETS...................73
CUSTOMIZING YOUR WIDGETS TIPS ............74

## CHAPTER 9: ACCESSING CONTROL CENTER.................................................... 76

Open Control Center ............................................76
Key Features Of Control Center ..........................76
   Bluetooth and Wi-Fi ......................................77
   Brightness and Volume Controls.....................77
Additional Options ............................................. 78
MASTERING YOUR DESKTOP WITH STAGE MANAGER ...................................................... 78
   What is Stage Manager? ................................. 78
   How to Use Stage Manager ............................79
Microphone Recording Indicator........................79
   Appearance .................................................... 80

Interpreting the Indicator .............................. 80
Manage Which Apps Can Access the Microphone ................................................... 80
Pinning Control Center Favorites ...................... 81
REMOVING ITEMS FROM MENU BAR .......... 82

## CHAPTER 10: WHAT IS SPOTLIGHT? .... 84

Opening Spotlight ............................................... 84
Searching For Items ........................................... 84
Opening Applications ......................................... 85
Performing Quick Actions .................................. 86
Tips For Effective Use Of Spotlight ................... 86
Currency, Temperature And Measurement Conversions ....................................................... 87
Narrowing Your Search ...................................... 89

## CHAPTER 11: USING SIRI SUGGESTIONS ............................................................................ 91

Setting Up Siri .................................................... 91
Turning on Siri .................................................... 94
Using Siri For Everyday Tasks ........................... 96
More Examples Of Siri Commands .................... 97
Drawbacks And Remarks ................................... 98
Troubleshooting Siri ........................................... 98

7

## CHAPTER 12: USING ICLOUD WITH YOUR MACBOOK .................................................. 100

Signing In and Setting Up iCloud ..................... 100
Configuring iCloud Features ............................ 100
Using Your MacBook with Other Devices ......... 102
Accessing iCloud Content on Your MacBook .... 104
   Setting Up iCloud on Your MacBook ............ 104
   Accessing Your iCloud Content ..................... 105
   iCloud+ Benefits ............................................. 106
   Syncing Across All Devices ............................ 108
Automatically save your desktop and documents to iCloud Drive with your MacBook ..................... 108
   1. How iCloud Drive Works with Desktop and Documents ................................................... 109
   2. How to Configure iCloud Drive for Desktop and Documents on Your MacBook ................ 110
   3. Accessing Your Files Across Devices ......... 110
   4. Benefits of Storing Files within iCloud Drive .................................................................... 111
   5. Apply Changes and Device Synchronization .................................................................... 111
   6. Manage Storage and iCloud+ .................... 112
   7. More on Using iCloud Drive ...................... 112

## CHAPTER 13: STORING AND SHARING PHOTOS WITH ICLOUD ON YOUR MACBOOK ................................................. 114

1. Setting Up iCloud Photos on Your MacBook  114
2. Benefits of Storing Your Photos in iCloud Photos ................................................................................. 115
3. Setting Up and Managing iCloud Shared Photo Library ............................................................................. 115
Inviting People and Sharing Photos................. 116
4. How to Use iCloud Shared Photo Library Features ............................................................................ 117
5. Managing iCloud Storage ............................... 117
6. Viewing Photos on All Devices ...................... 118
7. Privacy and Security with iCloud Photos ...... 118

## CHAPTER 14: ACCESS PURCHASES ANYWHERE WITH ICLOUD ON MACBOOK AND ENJOY ............................................................. 119

How to Access Purchases on All Devices .......... 119
Benefits of Purchasing with iCloud Syncing ..... 120
Using Find My Mac To Protect And Locate Your MacBook ............................................................ 121
Setting Up Find My Mac .................................... 121
Find My Mac Benefits........................................122
iCloud's Role In Unifying Your Apple Experience ............................................................................122

## CHAPTER 15: GIVING YOUR MACBOOK SOME PERSONALITY ............................ 123

How to Change the Wallpaper ......................... 123
Add Widgets to Your Desktop ......................... 124
Customize Your Desktop Environment ............ 126
Create A Memoji On Macos................................ 127

## CHAPTER 16: WORKING WITH MULTIPLE SPACES ON MAC ................................... 131

1. Understanding Mission Control ................ 131
2. Entering Mission Control ......................... 131
3. Creating a New Space ............................... 132
4. Switch Between Spaces ............................. 133
5. Working in a Space .................................... 133
6. Configuring Each Space ............................ 133
7. Managing Spaces....................................... 134
Different Ways to Switch Between Spaces On Mac ................................................................. 134
1. Trackpad Gestures .................................... 135
2. Using a Magic Mouse................................. 135
3. Using Keyboard Shortcuts ........................ 135
4. Using Mission Control .............................. 136
5. Using the Touch Bar (For MacBook Pro Models) ........................................................... 137
Tips For Spaces Use......................................... 137

How to Assign Apps to Spaces on macOS Step by Step .................................................................. 138
  Assign To Options: ........................................... 139
  More Tips To Manage App Behavior In Spaces 140
  Practical Applications of Assigning Apps To Spaces .................................................................................. 141
  Moving An Application Window From one space to the other .......................................................... 142
    Method 1: Dragging to the Edge of the Screen .................................................................................. 143
    Method 2: Moving Windows via Mission Control .................................................................................. 143
  Deleting A Space ............................................... 144
  Exiting Full-Screen Or Split View In Spaces ..... 145
  Best Practices For Managing Spaces ................. 146

## CHAPTER 17: SYSTEM SETTINGS ....... 148

  Accessing System Settings .................................. 148
  Modifying Key Settings ....................................... 149
  Keeping Software Up To Date ........................... 150
  Finding Specific Settings Quickly ..................... 150
  Locking Your Screen .......................................... 150
  Choosing And Customizing A Screen Saver ...... 151
  Customizing Control Center And Menu Bar ..... 151
  Updating Macos ................................................. 152

Icloud And Family Sharing Settings .................. 153
Adjusting True Tone For Ambient Light........... 154
Setting Up Dynamic Desktop For Time-Based Visual Changes .............................................................. 154
Changing How Items Appear On The Screen ... 155
Enabling Dark Mode To Stay Focused............... 156
Setting Up Night Shift For Warmer Colors....... 157
Connecting An External Display ....................... 158

## CHAPTER 18: ACCESSIBILITY SETTINGS ................................................................................. 161

OPENING ACCESSIBILITY SETTINGS........... 161
ACCESSIBILITY CATEGORIES AND FEATURES ................................................................................. 161
   1. Vision ....................................................................... 162
   2. Hearing .................................................................. 163
   3. Mobility ................................................................. 164
   4. Speech .................................................................... 165
Tips For Customizing Accessibility Settings..... 166

## CHAPTER 19: ADJUSTING YOUR VIEW WITH ZOOM ............................................... 167

How to Set Up Zoom on Your MacBook ........... 167
Key Features and Options of Zoom.................... 167
Voiceover – A Screen Reader For MacOS.........168

12

How to Activate VoiceOver (or disable it).....169
Key VoiceOver Features and Usage...............169
Zoom In On Words With Hover Text................ 171
Configuring Hover Text................................. 171
Using Hover Text............................................ 171
Customizing Hover Text................................. 172
Adjusting Your Mac Display Colors With Color Filters...................................................................... 172
Set Up Color Filters ........................................ 173
Select and Configure Color Filters.................. 173
Color Filters Quick Toggle ............................. 173

## CHAPTER 20: HEARING ACCESSIBILITY FEATURES – LIVE CAPTIONS .............. 175

How to Enable Live Captions............................ 175
Some Key Features of Live Captions................. 175
Limitations of Live Captions ............................ 176
Pair Hearing Devices With Your Mac ............... 176
How to Pair Hearing Devices ......................... 177
Once Paired:.................................................... 177
Customizing Airpods For Hearing Assistance .. 177
Backgrounds Sounds To Mask Ambient Noise. 179

## CHAPTER 21: VOICE CONTROL ON MACBOOK ................................................. 181

13

Personalize Your Pointer For Easier Viewing...182

3. Better Keyboard Access For Full Control......183

Have your Mac Speak for You with Live Speech ................................................................................184

Use Voice Shortcuts For Instant Actions..........186

How to Configure Voice Shortcut..................186

Let Siri Listen Out For Atypical Speech............187

## CHAPTER 22: MANAGING WINDOWS.. 189

Preview ...............................................................189

Full-Screen Mode For Focus On A Single App ................................................................................189

Split View For Multi-Tasking With Two Apps ................................................................................190

Stage Manager For Automatic Window Management ................................................... 191

Mission Control For Quick Access To All Open Windows ....................................................... 191

Multiple Desktop Spaces (Virtual Desktops) 192

Using Full-Screen View On MacBook...............193

Using Split View On MacBook .........................194

Full Screen Vs Split View: ................................196

Tiling Windows On MacBook .........................196

More Window Management Tips....................198

Using Split View For Extend .............................199

Widgets On Your MacBook Desktop ................. 199
Stage Manager ................................................. 201
Managing Windows and Desktops on Your MacBook with Mission Control ........................ 203
Create And Manage Desktop Spaces ................ 205
    Deleting Spaces ............................................ 206
    Mission Control-Usage Benefits ................... 207
The Red, Yellow, And Green Buttons ............... 207

## CHAPTER 23: MAC OPERATING SYSTEM REINSTALLATION ..................................... 211

Step 1: Reinstallation - Preparation .................. 211
Step 2: Turnoff Your Mac ................................. 211
Step 3: Start From MacOS Recovery ................ 212
    Steps To Startup From MacOS Recovery On Apple Silicon ................................................. 213
    Step-By-Step Guide To Start MacOS Recovery On Intel-Based Macs ..................................... 215
Reinstallation Of MacOS From MacOS Recovery ........................................................................ 217
Important Considerations During Reinstallation ........................................................................ 219
Tips For Successful MacOS Recovery .............. 220

## INDEX ..................................................... 222

15

# CHAPTER 1: APPLE MACBOOK PRO GETS SPEED BOOST AND ANTI-GLARE SCREEN

The new 14" and 16" MacBook Pros utilize some of the most robust Apple silicon yet, in the M4 Pro and M4 Max processors. One week after introducing a new MacBook and Mac Mini, Apple introduced its latest updated 14- and 16-inch MacBook Pros, with the M4 chipset platform. These can be upgraded to the game-changing M4 Max processor, which is a tremendous uptick from the last models back in the previous year. Much of the updates are incremental, but it gave the hardware even bigger power and performance capabilities.

## PRICE AND CONFIGURATION OPTIONS OF MACBOOK PRO 2024

The base model of the 14" MacBook Pro begins at $1,599, with essential features and specifications that work best for general usage. But for power users, there's also the ability to upgrade most components, and it could go as high as $3,199 with a higher core count on the CPU and GPU. Beyond this point, one can maximize storage and unified memory to make it a very powerful machine, be it video editing, 3D rendering, or software development.

In contrast, the 16-inch MacBook Pro begins at $2,499. This model is directed towards pros who want to see a larger screen and are hungry for more powerful specifications. If you do a full upgrade, the price touches $3,999. Similar to the 14-inch model, you can have more CPU and GPU core counts, increased storage, and higher unified memory capacity.

Both models begin shipping on November 8. This will put Apple in a good place to cover the needs of everyone, from casual consumers to high-end professionals, with a wide enough range to meet or exceed performance requirements and budgets.

## NEW CHIPS IN TOWN: IMPROVED DISPLAY AND CAMERA FEATURES

The MacBook Pro 2024 remains a marvel with its current set of features, with some major boosts. With the same **Mini-LED screens** available on the predecessor model, this laptop can provide excellent contrast and saturation of colors, something really important for creative professionals to take a look at. The **120-Hz refresh rate** allows smooth scrolling and responsiveness of the computer, even easier and more enjoyable to work with in jobs that require animation or game playing.

On the camera side, the same 1080p webcam takes center stage, but with this version bumping up to a 12-megapixel webcam, that adds Center Stage. That's a feature that auto-keeps you in the center of the frame during video conferencing, keeping your image more interesting and professional-looking. Until now, Center Stage had been available only on iPads and Studio Display, but it arrives for MacBooks, which can dynamically adapt to improve video conferencing. This has been very helpful, especially with employees working remotely and content creators who would be constantly participating in video communications. It's not that the audio experience has been lagging - the six speaker sound system gives premium class to deliver immersive audio quality, making it perfect for everything from music production to movie watching.

Besides that, Apple has fitted in the nano-texture glass option, just like the Studio Display, for much less glare. This comes in handy when someone works near a window or outside because it cuts down on reflections and distractions. The catch, though, is that this is an optional upgrade-an additional $150- and it's available across all MacBook Pro configurations.

## ADVANCED CONNECTIVITY OPTIONS AND WAYS OF SETTING UP YOUR MACBOOK PRO 2024

The MacBook Pro 2024 features a much more robust set of connectivity options, increasing its versatility for the wide range of user workflows. For instance, it now has three USB-C with Thunderbolt 4 along the bottom of the base model line, versus just two from its predecessor, the M3. This offers much better connectivity options with users able to connect several high-speed peripherals simultaneously for displays, storage, and much more.

Apart from USB-C, it also includes an HDMI port for easy and direct connections to external monitors and TVs without the need to purchase additional adapters. This could be of especial value for working professionals who often have presentations or need multi-screen setups. The inclusion of the SD card slot further makes the MacBook Pro appealing for photographers and videographers, as media file transfer from cameras can be done more conveniently with it. It also provides a high-impedance headphone jack-a great addition to audiophiles and professionals alike, seeking high-quality output audio. The other main highlight is in the inclusion of a MagSafe charging port, which users have come to love for its secure and easy connections while charging. This will definitely enhance convenience besides reducing the risk of accidental disconnections.

Sleek in space black and silver-elegant finishes, the MacBook Pro lets each user choose further how they want to show their style.

All three Thunderbolt 5 ports (USB-C) are available with the higher-performance configuration options on models with the M4 Pro and Max chips. This extension allows for data transfer at an even higher speed, necessary for professionals who deal with large files. For example, video editors and graphics designers need to have efficient workflows.

While one of the strong points of the 14" MacBook Pro is the fact that buyers can configure it with the base-level M4 chip, along with unified memory of 16GB, the same is not true of the 16-inch model. If you want to buy the bigger MacBook, you'll need to configure it with either the M4 Pro or M4 Max to get the power required for more resource-intensive applications.

## COMPARING THE M4 PRO AND M4 MAX: PERFORMANCE AND TECHNOLOGY ADVANCEMENTS

The MacBook Pro 2024 includes two chipsets: the M4 Pro and M4 Max, both positioned for various performance needs. Understanding the differences between these two options will be key for users wanting to optimize their workflows based on particular requirements.

Starting with M4 Pro chipset, it can be adjusted to sport up to a 14 to 20 core CPU. With such a setup, really power-hungry applications - video editing, 3D modeling, and software development, among others - become pretty apt. The M4 Pro finds a nice balance between power and efficiency for the professional who needs robust capabilities but does not have to opt for the highest of high-end hardware.

On the other hand, the M4 Max is the powerhouse of the lineup. The base model comes with a 16-core CPU, up to an astonishing 40-core GPU for almost unparalleled performance - perfect for the most demanding applications: high-resolution video editing, complex graphic rendering, even heavy data processing. The M4 Max is designed to handle the most intense workloads with ease; it's all about giving users the speed and power they need for professional-grade applications.

Both chipsets are manufactured using a second-generation 3nm process that gives way to a larger amount of transistors inside a smaller space. Such furtherance guarantees increased efficiency and speed, with chips executing more operations per second and consuming less power. Long-time battery life, better thermal management, and an overall seamless user experience all finally come out from such efficiency.

Along with raw performance, the M4 Pro and M4 Max feature advanced graphics such as mesh shading together with ray tracing. According to Apple, the ray-tracing engine in these chipsets is now running at double the speed compared to their predecessors, the M3 chips, driving realistic lighting and reflections for the most graphically intense applications. This surely serves as a good boost for game developers and visual effects artists who rely on refined graphics to suck users into virtual worlds.

Second, Apple Intelligence that powers many features in macOS 15.1 includes both chipsets. In this regard, it has a suite of artificially intelligent features to make productivity easier. Amongst the major features included are Writing Tools to assist with content creation, transcription summaries in the Notes app, the Clean Up feature in Apple Photos, and Smart Replies across Messages. These smart tools ease daily tasks and improve interaction with the operating system in general, as it enhances its user-oriented capability not only for professional users but also for everyday personal uses.

## UPGRADE TO THE MACBOOK AIR AND A REFRESHED MAC LINEUP IN 2024

The new MacBook Air now comes with 16 GB of RAM as the base, compared to 8 GB, in a tactical move by Apple. At $999, this tweak may be so that the MacBook Air can cope better with tasks associated with Apple Intelligence features such as Siri, Dictation, and Enhanced Dictation, that tend to be voracious for more system resources. With increased base memory, Apple ensures that users enjoy smoother performance, especially when using AI-powered productivity enhancements and professional applications that have become core parts of working in macOS.

Introducing 16 GB of RAM across the board of MacBook Air is an indication that Apple has now become serious about user experience across the line. Coming off general improvements in the MacBook Pro 2024, this further epitomizes Apple's concern to provide powerful devices, not only capable of handling advanced features but also with smooth integration.

By updating the MacBook Air at the start of the year, almost the entire Mac line-up has seen a refresh in 2024. This spate of hardware announcements shows Apple's intention to modernize its products and stay competitive in the consumer tech world. The refreshed lineup now comprises the MacBook Pro, powered with advanced M4 Pro and M4 Max chipsets to ensure that professionals and creatives get powerful machines that address their demanding workflows.

The only remaining devices in this update wave are the Mac Studio and Mac Pro, currently touting the M2 Ultra chip. That chip is unprecedented, because it's able to offer a GPU with 60 or 76 cores for unmatched graphics performance. M2 Ultra is created to handle resource-intensive tasks and is the most powerful chip Apple has in store for professionals who demand maximum graphic performance and memory bandwidth.

Looking ahead, it's rumored that the Mac Studio as well as Mac Pro will migrate to the M4 platform starting 2025. That could further increase configurations and capabilities, and at least better set Apple's positioning in high-performance computing. Fundamentally, the 2024 refresh of the Mac line-up, led by an upgraded MacBook Air and a new MacBook Pro, serves as testimony to Apple's commitment in providing front-line technology that addresses the emerging needs of its patrons. All memory upgrades on top of power chipsets ensure the Apple devices remain at the very top of performance in AI-intensive and other creative applications.

# CHAPTER 2: SETTING UP YOUR NEW MACBOOK

When setting up a new MacBook, it is important to go through each setup option carefully in order for the configuration to be just right. The setup process in and of itself is not overly complicated; however, it does take some time-particularly if you plan on reviewing and adjusting each available setting.

These steps are good for any new Mac. Note that for this tutorial, we assume you have a brand new MacBook and are setting it up from scratch.
To start working on a new Mac, you will want to set this up with the appropriate settings so that it's ready to use. Generally, setup is quite easy and doesn't take much time; however, it can if you want to go through each setting to turn some options on and off or change them.

## SET LOCATION AND ACCESSIBILITY SETTINGS

After you power up your MacBook, select your country or region. You can then turn on the features in accessibility to help you if you have problems with eyesight, hearing, and other items. If you want to set any of these, tap accordingly. Otherwise, you may tap "Not Now" to continue and later set them in case you may require them.

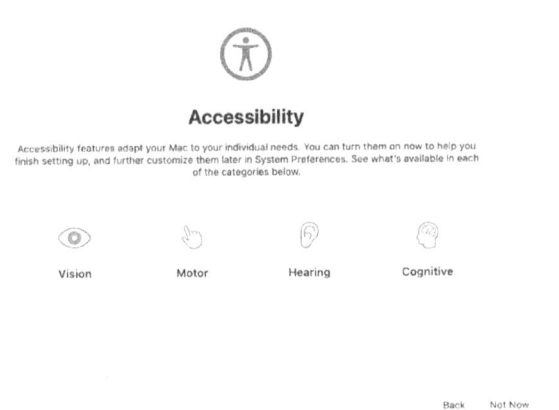

## CONNECT TO WI-FI

Next, your MacBook will connect you to a Wi-Fi network. Select the network you want, enter the password, and tap Continue. You will then proceed to a page highlighting the privacy policy regarding your data. Tap the link Learn More if you want more; otherwise, tap Continue and proceed.

## TRANSFER DATA WITH MIGRATION ASSISTANT

You will now be given the opportunity to use Migration Assistant-a utility that transfers data from another Mac, a Time Machine backup, or a Windows PC. If you have any content to migrate, you can select the correct option and follow the on-screen instructions to complete the transfer. If you want to skip this step, tap "Not Now." You can still use Migration Assistant to transfer data later-even after you've completely set up your MacBook.

**Migration Assistant**

If you have information on another Mac or a Windows PC, you can transfer it to this Mac. You can also transfer information from a Time Machine backup or another startup disk.

How do you want to transfer your information?

○ From a Mac, Time Machine backup or Startup disk
  From a Windows PC

Not Now          Back  Continue

## SET UP APPLE ID AND LOGIN

Sign in next with your Apple ID or create a new one if you don't have an account yet. An Apple ID is required to access most of the Apple services. If two-factor authentication is enabled for the Apple ID, enter the verification code sent to the iPhone or iPad.

Click Continue and then review the terms and conditions for iCloud.

## CREATE A COMPUTER ACCOUNT

At the Computer Account screen, first confirm your full name and create an account name that you will be using to login into your Mac. Then, Type your password for the account and a password hint if you want so you might not forget it later. Put the check alongside "Allow my Apple ID to reset this password" to make account recovery easier should you forget your password.

To personalize your account even further, tap the image alongside your account name to choose a Memoji, emoji, monogram, or photo. Tap Continue once you have customized your account to your satisfaction.

## KNOW THE FIND MY APP

The next screen provides you with some information about the Find My app. This is helpful because it serves to let you locate your Mac if it becomes lost. Taking a minute to understand how this works can also give you peace of mind, knowing that should this ever happen to you, you have ways of tracking your device.

## MAKE THIS YOUR NEW MAC

You will be taken to the "Make This Your New Mac" screen where you can go through and change a number of different settings. You may click Continue to take all defaults for all the settings or at the bottom, you can click Customize Settings to change some options for Location Services, Analytics, and Siri.

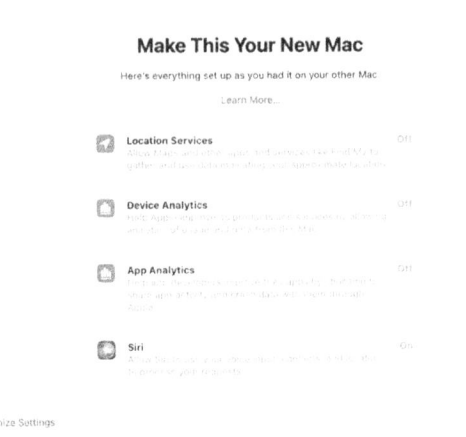

Under Location Services, you are able to enable/disable location-based apps: Apple Maps and Weather.

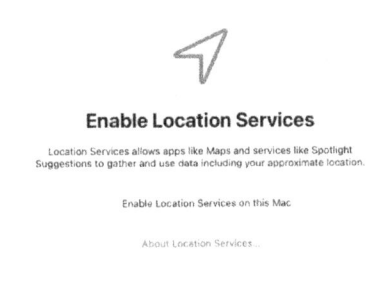

You also have to select your time zone: you can either choose it from the drop-down list or click on a place in the map.

Next, you want to decide if you will share analytics data with Apple. If you want further information on how data are collected and used, click the About Analytics and Privacy link.

Finally, if you'd like to be able to limit which apps and which websites have access to your MacBook, you might want to enable **Screen Time**. To configure this later you can click **Set Up Later**.

# ENABLE AND CONFIGURE SIRI

In the opened window, check the box labeled Enable Siri. This will turn on Apple's voice assistant for your MacBook and provide a hands-free way to interact with your computer. Once you flip that switch to enable Siri, you're taken through a series of screens where you're asked to say certain phrases, because Siri wants to get a little more familiar with your voice.

**Siri**

Siri helps you get things done just by asking. Siri can also make suggestions before you ask in apps, search, and keyboards.

☑ Enable Ask Siri

About Ask Siri, Dictation & Privacy.

Back    Continue

If it doesn't bother you that Apple will have access to your recordings to help improve Siri, check the box next to **Share Audio Recordings**. The data enables Apple's voice recognition to fine tune its service and also aids in improving Siri in general. Tap the About Improve Siri and Dictation & Privacy link for additional details about how your recorded audio is used and stored.

Of course, if you don't want to do that, then select Not Now. That way you keep all of your audio recordings private, though you'll still be able to use Siri for basic functions. Of course you can always go back into Settings later and turn this on.

## SET UP FILEVAULT FOR DISK ENCRYPTION

Once Siri is set up, you will be asked whether you want to turn on FileVault disk encryption on your MacBook. FileVault is a security option that encrypts your entire disk so data is kept safe and resistant to unauthorized access. To turn this extra layer of security on, select **Turn on FileVault disk encryption**.

### FileVault Disk Encryption

FileVault secures your data by encrypting the contents of your disk and locking your screen with a password.

Would you like to use FileVault to encrypt the disk on your Mac?

☑ Turn on FileVault disk encryption
☑ Allow my iCloud account to unlock my disk

Your iCloud account "lance@lancewhit.com" can be used to unlock your disk and reset your password if you forget it. If you do not want to allow your iCloud account to reset your password, you can create a recovery key and store it in a safe place to unlock your disk.

Back   Continue

You can further enhance your security by selecting the box beside Allow my iCloud account to unlock my disk. If you can't recall your password on Mac, you may be relieved to know this method will afford you another way to access your encrypted disk without manually going through some kind of reset process.

## SET UP TOUCH ID (IF AVAILABLE)

If your MacBook includes Touch ID, you should set that up next. Click **Continue** and follow the onscreen prompts to enroll your fingerprint. You will have to place and remove your finger a couple of times onto the power button until your fingerprint is registered. That way, Touch ID will be able to recognize your finger for its safe unlocking later. When Touch ID is ready, click **Continue**.

## CHOOSE A DESKTOP THEME

Next, you'll have the option to choose a theme for your desktop after setting up Touch ID: Light Mode,

Dark Mode, or Auto. Light Mode is bright and shiny, while Dark Mode is sleek and much easier on the eyes in low-lit environments. The Auto option will change the themes according to your time of day, seamlessly transitioning between light and dark mode.

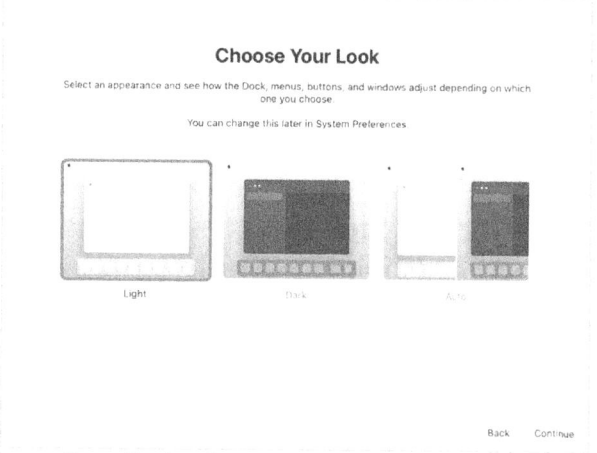

**Choose Your Look**

Select an appearance and see how the Dock, menus, buttons, and windows adjust depending on which one you choose.

You can change this later in System Preferences.

Light    Dark    Auto

Back    Continue

Even after the default setting when you first set up your system, later in life, you can change how you prefer your theme through the Display settings. This will give you flexibility in order to personalize your visual experience due to personal preference or certain lighting conditions.

## UPDATE YOUR MAC

Once this is done, you'll be whisked away to the desktop of your newly set-up MacBook. The final important thing you'll want to do is update your Mac- you want the most recent version of macOS for many reasons, one of which is new features. Also, critical patches and bug fixes are added.

The first thing to do to update is to open the System Settings: Click the Apple menu located at the top left of your screen. Next, click General, followed by Software Update. It will check under this section for any updates. If there were any available updates, it may ask for your password for your Mac. Just insert the password and click Update or Install, and it will let it download and install itself.

Those using Macs with macOS Monterey or an operating system earlier than this will have the steps slightly different. Click the Apple icon located in the top left, and then click About This Mac. Under this menu choice, select Software Update; you will be taken to the update screen as described above. You will see if there is an update available. Click Upgrade Now to start installing.

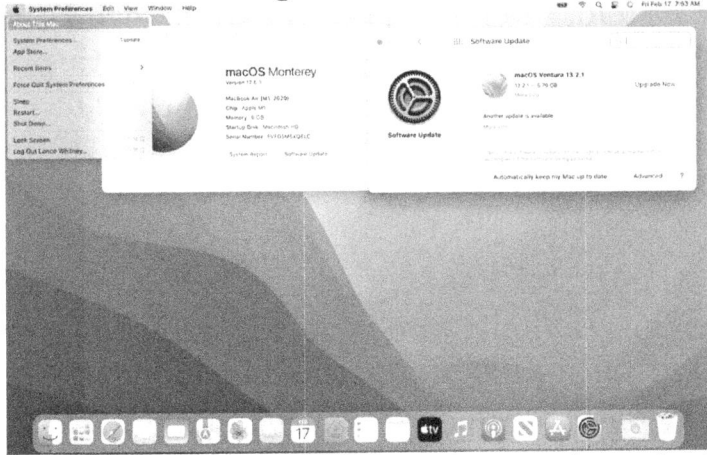

This may take some time after having launched the update process, as your Mac needs to download the files to do so. The length of time depends on the update size and your Internet connection. Once the files have finished downloading, installation will begin.

During this, your Mac will restart a few times. When it finally comes to a stop after the last reboot, it returns you to the login page from whence you will have to log in with your credentials. You'll log into a now-updated MacBook with the latest software enhancements that guarantee a much easier and more secure usage experience.

Watching for updates will keep your Mac current and support all the improvements Apple makes to the OS. This last step will ensure not only that your new MacBook is fully prepared to run at optimal levels but that you too are prepared with new, updated features and safety enhancements.

# CHAPTER 3: NAVIGATING THE MAC OS INTERFACE

## UNDERSTANDING THE LAYOUT: THE DESKTOP AND THE MENU BAR

When you first power up your MacBook, you are greeted with the desktop. The desktop is the central workspace where you will spend most of your time working. Along the top is the menu bar, which houses all of the system functionality and application menus. Along the bottom you'll find the Dock, where you can access quick ways to get into your most-used applications and documents.

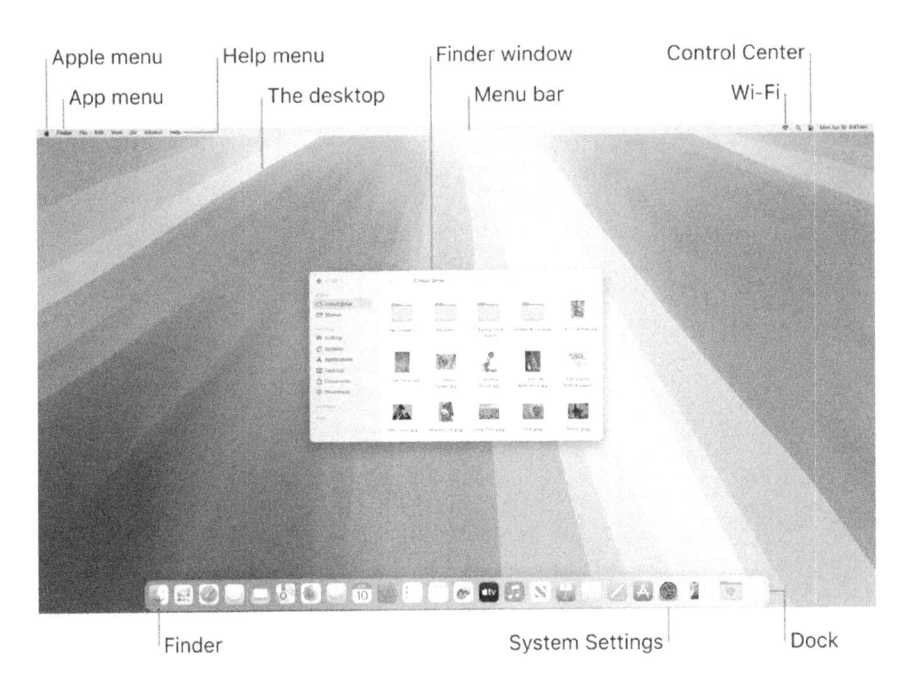

**Hint**: If you are having trouble finding where your cursor is on the screen it can be temporarily magnified by quickly moving your finger back and forth on the trackpad. If using a mouse, try sliding it back and forth quickly.

## DESKTOP FUNCTIONALITY

The desktop is the home base for your work. It's where you can open and work in applications, work with your files, and use Spotlight to search your MacBook or the web. You can customize your desktop by changing your wallpaper. To do this, open System Settings, click Wallpaper inside the sidebar, and then select how you want to go about it.

You can also store files right on the desktop if that is how you roll. If you want your desktop to remain somewhat organized, you can use the stacks option, which lets you group similar files together, keeping clutter at bay.

## THE MENU BAR

Along the top of your screen is the menu bar; this is a key element of macOS. It is the control room through which one can access other functions to carry out certain tasks inside applications. On the left-hand side, menus are context sensitive: the items that appear will change depending on what application you happen to have open at any one time. This adaptability allows for an efficient workflow in which commands and tools relevant to the opened application can be easily accessed by the user.

On the menu bar's right, there is a series of icons offering quick ways to access some of the important features. These include the following :

1) Wi-Fi Status: The Wi-Fi icon 🛜 in the menu bar allows you to join a network or display information about your current connection.

2) Control Center: This icon gives access to settings like brightness, volume, and Do Not Disturb.

3) Battery Charge: This icon 🔋 allows Mac users to easily see how much battery life they have remaining and opens up options to control your power settings.

4) Spotlight Search: The magnifying glass icon 🔍 is an easy way to search your Mac and the web to find whatever documents, applications, or anything you may need in an instant.

Using these menu items listed in the menu bar, you will be able to handle your tasks with ease and keep your productivity in good condition.

## THE APPLE MENU

The Apple menu  is important to the macOS experience and is found on the left-hand side of the screen. You can always open the Apple menu by clicking the Apple icon. It's where you'll find many of the frequently used items that create an easy-to-access link to several key functions.

In the Apple menu , you will be able to select:
1) About This Mac: See information about your Mac, including the model and which version of macOS you're using.
2) System Preferences: This opens the system preferences with which you can make changes to features and functionalities on your Mac.
3) Recent Items: This opens a list of all applications and documents that you have opened lately and offers quick access to the documents you have been working with.
4) Shut Down, Restart or Sleep: Just put your computer to sleep or restart/shut it down from here.

The Apple menu  allows you to access some important system functionalities, adding convenience for you when you navigate through macOS.

## APP MENU

One of the powerful features of macOS is the ability to run multiple applications and windows simultaneously. At any given time, the name of the active application is boldly type-faced to the right of the Apple menu  in the menu bar. To the right of the app name you will find its unique menus, which provide access to the functions and commands specific to that application.

If you switch to another application or click on an open window of another app, the name in the app menu changes. That means the menu options within the menu bar will change with the commands relevant to the newly active app. This dynamic updating ensures you continuously have access to the relevant tool set you need for your current task.

If you have opened a menu looking for a certain command and do not find it, remember to check whether the application in which you want to execute the command is the active application. If it is not, just click in the desired application window to bring it to the front, and the application menu bar will change to reflect the appropriate application commands .

## THE HELP MENU

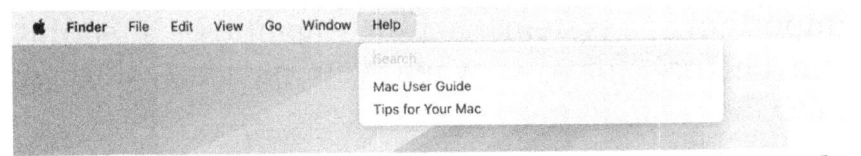

macOS provides great help options via the Help menu, which is always available via the menu bar. Using this feature, you can find your way around your MacBook at any moment. In opening the help, in opening the Finder by clicking on it from the Dock. While in this environment, click on the Help menu and select macOS Help to open the macOS User Guide. The user guide is quite comprehensive and gives one a chance to go through troubleshooting and locating most of the operating system's features.

Alternatively, from any page in the Help menu, you can simply type your search term into the search box and click Return. As you start to type, suggestions will begin to show, making it easier to find what you want.

To use application-specific help, open an application, then from the menu bar click Help. This will give you context sensitive help so you can search for application-specific solutions related to the work you are trying to complete.

Tip: Maximize your desktop by making use of the widgets in your widget gallery. You can even share widgets with your desktop directly from your iPhone, and without having to download the corresponding app, the information needed will seamlessly appear at your fingertips.

# CHAPTER 4: THE FINDER

Finder is one of the most important parts of macOS.

It is represented by a blue icon with a smiling face , and it's where you do the bulk of organizing and finding almost anything on your Mac, including documents, images, movies, and many other types of files. With Finder, you can manage your files and folders efficiently to continue with smooth workflow.

## OPENING A FINDER WINDOW

Click the Finder icon, which appears in the Dock at the bottom of your screen. The Finder window opens, displaying a sidebar that provides easy access to your key locations, including Documents, Downloads, and Pictures. In this layout, you will find a way of navigating around your files.

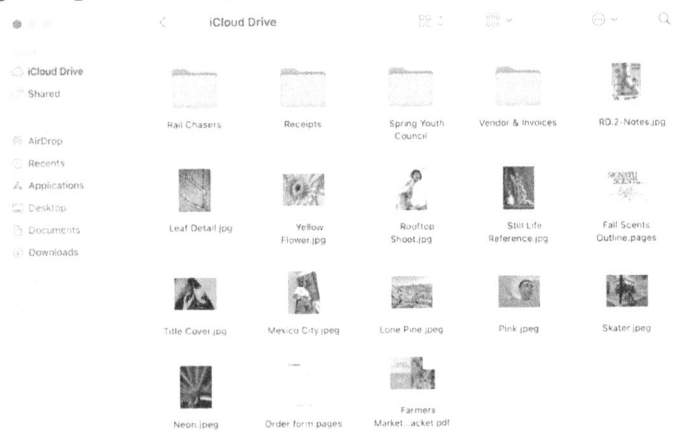

## QUICK VIEW AND EDITING FILES

Another of the more useful features of the Finder is that file icons respond to the Force Click function. By pressing firmly on a file icon, you pop open a quick view of its contents with it opening the file fully. That comes in pretty useful for identifying what might be inside a document or image. Also, if you want to edit the name of a file, by Force Clicking on the filename it becomes editable so that you can rename the file faster.

## ORGANIZING

Your Mac has already created default folders for commonly used types of content: Documents, Pictures, Applications and Music. By default, these folders will automatically categorize your files. As you progress through writing documents, installing apps, or doing other work you may want to create additional folders for further organization.

To create a new folder, starting from the top of the screen, go into the Finder menu option File > New Folder. This action will create a new folder in the current location you are looking at and name it appropriately for related files with which you will store them. This organizational strategy helps maintain a clutter-free workspace and makes it easier to find your files when needed.

## SYNCING DEVICES

When you connect a device to your Mac-say, an iPhone or an iPad-it becomes available right inside of

Finder. When attached, the integration of such a device becomes effortless: the name will automatically appear in the Finder sidebar.

All you have to do is click on the name of the device in your sidebar to manage your device. An interface opens that shows a number of options through which you can keep your device maintained. Some of the major tasks that you can do are:

1) Backup - Taking a backup of your device so that your information is kept safe, and you get it back when you need it.
2) Update: The device will automatically check for any software updates of the device and will keep it running with the most recent features and security enhancements.
3) Sync: It synchronizes the device with the Mac, thus allowing the transfer of files, music, photos, and other content between these two devices.
4) Restore: Whenever necessary, return your device to the default settings or even to an earlier backup that may be helpful if you have any problems.

This makes it easy to keep your devices in sync and up to date without having to leave the Finder.

## THE FINDER SIDEBAR

The Finder sidebar is one of the most important elements that will make the use of macOS much easier. On the left side of the Finder window is a sidebar that shows items you use regularly or want easy access to. This keeps everything nice and neatly organized for ease of file management.

Some of the important things that it usually contains in its sidebar include:

1) iCloud Drive: Once you click the iCloud Drive folder, you can see all of your documents that are stored in iCloud. Thus, it allows you to access those files kept in the cloud, making it very easy for you to work on a document from any of your other devices.
2) Shared Folder: This folder will show you documents you are sharing with others, as well as documents others are sharing with you. This is the best way to handle collaborative projects or access certain files from coworkers without having to search your entire file system.

To customize what appears in the sidebar, go to the top of the screen and select **Finder** > **Settings** from the menu. From the settings menu, you have the option to toggle on and off which items you want showing up in the sidebar, so that you can tailor it to your workflow and preference. This will keep some of your most used files and folders just a click away.

## ADJUSTING FILE AND FOLDER VIEWS

macOS offers several methods to view files and folders in the Finder that you can use to select whichever you find most convenient and suitable for your needs. To change how you are viewing your documents and folders, locate the pop-up menu button at the top of the Finder window. The button opens a pop-up menu that allows you to choose any of several views:

1) Icon View ⊞: This view presents your files and folders as icons. It is an intuitive layout that immediately catches your eye, making it great for fast identification of files based on their visual appearance. Icon view is rather useful for image-containing folders, where icons will display thumbnails of your pictures.

2) List View ☰: This is a view in which the files and folders appear as a list. There are other details concerning this view, like the file size, date modified, and file type. In this kind of layout, it would be easy to sort out and manage your files, especially if you were dealing with many document files. Sorting of files in accordance with different attributes can be done by clicking on column headers.

3) Column View ⫞: This is a tree-like view that displays the contents of each folder in a separate column as one navigates through folders. So each time you select a folder, its contents open in another column. It's easy to drill into your directory structure and not get lost.

4) Gallery View ▭: This is a visually rich layout showing an enlarged preview of the chosen file. For this reason, this view is perfect for detecting images and video clips, as well as other documents, very fast. From here, in the Gallery View you may also carry out all those fast changes

in the file, like rotate or mark up an image right within its preview.

## USING THE PREVIEW PANE

Of course, beyond just simple handling of files, there's one preference that you can turn on in Finder to make using it a bit easier: the ability to display the Preview pane. To do this, go under the menu bar and select View > Show Preview. By doing so, you will turn on a right-side panel in the Finder window that will provide additional information about the currently selected file.

To customize what displays in the Preview pane, based on the type of file you are working with, select View > Show Preview Options. You will now be able to select what options to display, saving you time by not having to open the actual file.

## GALLERY VIEW

If you like to have the filename on the side of their respective files in Gallery View, turn this setting on. Press Command-J. This opens the view options. Then check the checkbox for Show filename, which adds the filenames underneath the previews. You still get to take advantage of having big visuals but can quickly identify what your files are.

## QUICK ACTIONS

macOS introduced a feature with Gallery View and Column View called Quick Actions. These permit you to manage and edit files in the Finder without opening other applications. To access Quick Actions, click the button ⋯ at the bottom right of the Finder window.

Quick Actions give you access to a number of shortcuts which go a long way towards smoothing your productivity and workflow. A few of the things you can do include:

1) Rotate Images: It enables you to rotate images to your preferred orientation without necessarily opening an image editor.
2) Markup and Crop Images: You can markup or crop images right in the Finder by using the Markup tool. That's really useful when you just have a couple of quick edits you need to make.
3) You can combine the files; merging several images and PDFs into one single file. The documents or

presentations you will create will be well-organized.
4) Trim Audio and Video: You do have the capability with multimedia to trim audio and video without having to actually open an editing application; therefore, you have the capability of removing sections from the video or audio that you don't need.
5) Run Shortcuts: Get integrated productivity by running custom shortcuts you created in the Shortcuts app. That said, this feature gives you the ability to automate repetitive tasks right from Finder.
6) Automator Workflows: For power users, actions can also be created using Automator workflows. You might create an workflow that watermarks images or performs other bulk actions to files to help automate a process for you.

## QUICK LOOK

Other powerful features in macOS include Quick Look. It allows you to view the contents of a file without having to open it completely. All you need to do to activate Quick Look is select a file in the Finder and press the Spacebar. You will instantly have a window open showing you what's inside the file.

In the Quick Look window's top, you have some buttons that you can click to:
1) Sign PDFs: While looking at a PDF, you can sign it right in the Quick Look interface without wasting extra time.

2) Trim Media Files: You can also trim audio and video files to your favorite length without leaving the Quick Look interface to do so.
3) Markup, Rotate and Crop Images: As in Quick Actions you can make basic adjustments to images - rotate or crop for example - directly in Quick Look.

## ACCESSIBILITY TIP

If you have images and want to add an alternative description, you may easily do so; thus, making them much more accessible. Using Markup with either Preview or Quick Look, one can add a description-a much-needed feature that VoiceOver will be able to read for those users dependent on screen readers.

## GET THERE FASTER WITH THE GO MENU

Finding your way into certain folders takes a number of clicks, but the Finder bar's **Go menu** can save you this hassle by easily accessing folders and locations. Here's how to make the most of it:
1) To get into common folders, like Utilities, all you have to do is go into the Go menu and click **Go > Utilities**. You won't need to click through several directories.
2) If you want to jump to the topmost level of nested folders, from here you can quickly go up along the folder hierarchy using Go > Enclosing Folder.
3) For a specific folder path, if you know exactly where you want to go, use Go > Go to Folder, then

simply type in the path. This allows you to go right to your destination in just a few keystrokes.

# CHAPTER 5: THE DOCK ON YOUR MAC

The Dock is one of the most recognizable features in the Mac operating system, and it appears by default along the bottom of your screen. It's a good place to store those applications and documents you use most to help you work more efficiently and effectively. Here's a step-by-step guide on using the Dock to launch, switch between, and manage your applications and files.

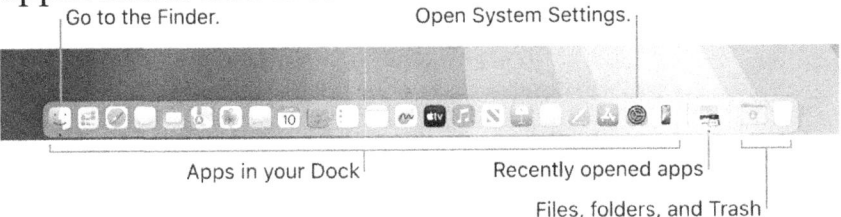

Go to the Finder.    Open System Settings.

Apps in your Dock    Recently opened apps
Files, folders, and Trash

## HOW TO OPEN AN APPLICATION OR OPEN A FILE

The features of the Dock are that it provides quick access to applications; hence, it's easy to launch what you need without browsing through unnecessary navigation via menus or Finder windows. There are a couple of ways to open an app or file from the Dock:

1) Direct Click: To open an application, click on its icon in the Dock: every icon represents an application that opens an application directly.

2) Launchpad : This is an icon in the bottom Dock; clicking on it will bring up a grid displaying

54

all applications installed on your Mac. You scroll through the icons and click to open any one of them.

3) Spotlight Search 🔍: For an even quicker way to find and open an app, use Spotlight Search. Located in the top right of the menu bar, click the magnifying glass icon or use the keyboard shortcut Command + Spacebar. Begin typing the name of the app you're looking for, and Spotlight will show relevant results. You can then select the app directly from the results to open it.

4) Recently Opened Apps: The center part of the Dock also shows your recently opened apps. This makes life even simpler because you can jump right back into applications you have opened recently, without searching for them.

## CLOSING AN APP

Knowing how to correctly close an application is very critical when you want to manage system resources and maintain high performance. Here's how you can effectively close an app:

1) Close a window: The red dot in the top-left corner of an opened window will close it, but the application itself won't close. It's good for cleaning up your workspace without quitting the app.

2) Open Apps: Those applications which are currently opened can be identified by a little black dot beneath their icons in the Dock. This will show

55

you that, though its window might be closed, the application is open.
3) Quit App: You quit an app to leave it completely shut and free up resources. To do this, control-click (or right-click) on its icon in the Dock. You see a context menu of options; choose Quit. This closes the application completely and removes it from active memory.

## ADDING AN ITEM TO THE DOCK

You can add items in your Dock in order to customize it and make your work easier:
1) Drag and Drop:
   a) Locate the application, file, or folder you want to place in the Dock.
   b) Click and drag the item, holding it toward the Dock
2) Keep the item where you want to reside
   a) Apps: Dock them in the left area of the Dock. It is designed for applications where you can have access to your frequently used programs
   b) Files or Folders: Drag and drop these in the right area of the Dock. This space is allocated for various documents, folders, and other forms of files.
3) Release Item: Once you have dragged the item into place, release your mouse button or trackpad to let it drop into the Dock. The item will sit there in the dock for easy access to help with efficiency in your work.

## REMOVING AN ITEM FROM THE DOCK

If you see that you no longer require the particular item in your dock then you can easily remove it:
1) Drag out of the Dock:
   a) Click and drag on the item you want to remove.
   b) Drag it out of the Dock and away from the icons.
   c) You will see a "Remove" label, indicating you can let go of the item.
2) Release the Item: When the item is outside the Dock, release the mouse button or trackpad. The item disappears from the Dock, but it isn't deleted from your Mac-the item is still available via Finder or Launchpad.

## MISSION CONTROL: SEE EVERYTHING OPEN ON YOUR MAC

Mission Control is a mighty feature that helps you to see all the applications, windows, and desktops currently opened on your Mac for effective workspace management. Here's how you get the best out of it: 1. Accessing Mission Control: a. Keyboard Shortcut: The default shortcut of Mission Control is F3 or the icon of three rectangles on the keyboard.
1) Trackpad Gesture: Perform a three-finger swipe upwards on the trackpad. This opens Mission Control, which gives a bird's-eye view of the applications and windows opened.
   1. Navigate Open Windows:

a) Mission Control gives you a view of all opened windows, full-screen apps, and desktop spaces in one place.
b) This feature allows you to click on any window or app to immediately jump into it, smoothing out your workflow and enhancing multitasking.
2. Add Mission Control to the Dock:
c) To have Mission Control easily at your fingertips, you'll want to add its icon to the Dock.
d) To do so open Mission Control, and then open the Applications folder. Locate the Mission Control icon ⊞, and drag it onto the Dock placing it where you like on the Dock.

## VIEWING ALL OPEN WINDOWS IN AN APP

Sometimes, you may have opened several windows in one application. Mac OS X makes it really easy to see what windows are open as well as switch between them:
1) Force Click on the App Icon:
   a) Locate in the Dock the app icon that has several windows opened.
   b) Force Click: This involves clicking on the app icon and pressing deeper in - or, in other terms, harder - on the trackpad or mouse. This will show you all the open windows of that application.

2) Switch Between Windows: Once you have this view, you can click any window to bring it forward to continue working with different windows without interruption.

## CUSTOMIZING THE DOCK

You can personalize the appearance of the Dock in the following manner with macOS:
1) Open System Settings: From the top left corner of your screen, click the Apple menu and select System Settings.
2) Click Desktop & Dock: In this section, you will be able to adjust several visual and behavioral settings related to the Dock:
   a) Size: This allows you to set the size bigger or smaller, depending on your preference. o Position: Move the Dock to the screen left or right side for a design you might find most appealing.
   b) Hide Option: You can just set the Dock to hide when it is not being used; this may give you a neater working environment.
3) Changing the Magnification/ Minimized Windows Behaviour: You can easily change the magnification effects that come along with determining how the minimized window acts/behaves.

## ADDITIONAL TIPS FOR USING THE DOCK

1) Changing Icons: The Dock is somewhat customizable, and one thing you can change is the icons on the Dock. Simply click and drag an app icon to the place in the Dock you want it to be. Release to set there.
2) Adding Applications: An application to be added to the Dock (in order to have fast access) must be located either in the Finder or Launchpad; afterward, click and drag it onto the Dock until it is dropped. Upon releasing it, the icon will stick to the Dock for easy opening the next time around.
3) Removing Applications: To remove an application from your Dock, click and drag the icon off of the Dock until you see a "Remove" label appear. Release the icon; it will disappear from the Dock without actually deleting the app itself.
4) Dock Preferences: You can open System Settings to alter the appearance and functionality of the Dock. You can change the size of the Dock, where it sits on the screen, and minimize window preferences within System Settings.

# CHAPTER 6: NOTIFICATION CENTER ON YOUR MAC

The Notification Center is a single place that centralizes all your notifications and information you may want to access, such as calendar events, stocks, weather updates, and much more. This will keep you updated and informed without cluttering up the desktop with a number of app windows.

## ACCESSING NOTIFICATION CENTER

You can access Notification Center by one of the following ways:
1) Clicking the Date or Time:
   a) The menu bar is usually located on the top right corner of your screen.
   b) Click the date or time that appears. Doing this will open the Notification Center where you will be able to see notifications and widgets.

2) Using a Trackpad Gesture: The more fluid gesture is the swiping left from the right edge of the trackpad with your two fingers. It opens the Notification Center, whereby you will have an easy and quick way to access notifications without getting out of your way of work.
3) Keyboard Shortcut: You can also use the F12 key or Control + Option + Command + D, depending on how your keyboard is set up to enable/disable the Notification Center.
4) Viewing Widgets: Once you open Notification Center, the top section is a list of notifications. Below these are several widgets that you can interact with. Scroll down to see more widgets and/or notifications as you would like to view.

## INTERACTING WITH NOTIFICATIONS

The Notification Center is more than just a collection of passively displayed alerts; it is interactive and allows for quick actions:
1) Read Notifications: Notifications are stacked based on application, showing you the most recent alerts first. You will notice icons identifying the type of notification, such as Mail, Calendar, and Messages.
2) Take action Many notifications will offer choices to take action directly:
   a) Allowing/Responding to Emails: Often when receiving an email through notification, you can reply within the notification itself without having to open the Mail application. Click the notification to respond or use quick reply options.
   b) Managing Calendar Events: Click on a calendar notification so that you can view details about upcoming events; you can often add or modify events right from this view.
   c) Podcasts: When you get an alert saying a podcast has arrived, you can click or tap on it to begin listening right away, or you may view more details about the episode.
1. Viewing More Options:
   a) Click the arrow at the top-right of a notification to expand it. More options open to you on that notification, such as extra actions, links, or related settings.

b) For example, one may reply or mark the conversation as read by expanding the message notification.
3. Using Notifications on your iPhone: As with iPhone Mirroring, notifications from the iPhone can also pop up on the Mac. You can interact with the notifications right from the Mac by using Multitasking across devices. If you receive a text message on your iPhone, it will pop up on your Mac, and you can respond from there.

## CUSTOMIZE NOTIFICATION PREFERENCES

You can further customize your Notification Center experience by adjusting notification settings for individual apps as follows:
1) Opening System Preferences: Click the Apple menu located at the top-left corner of your display, then select System Settings (or System Preferences, depending on your macOS version).
2) Selecting Notifications: Click on Notifications from the available choices. From here, you will have a list of all the apps that can show notifications.
3) Notification Settings: a. From the list, click any app to modify any settings related to notifications.
   a) Allow Notifications: ON/OFF to let/block notifications from that particular app.
   b) Alert Style: Choose how you want notifications to pop up in front of you, such as Banners, Alerts, or None.

c) Sounds: You can even set different sound options for receiving notifications from particular apps.
d) Notification Grouping: Choose whether notifications are kept grouped by apps or show up one by one.
4) Customizing Widgets: You can customize your Notification Center widgets. Where available, click the Edit Widgets button and drag-and-drop widgets to reorder them according to your needs. This allows you to keep the most important information at the top.

## SETTING YOUR NOTIFICATION SETTINGS

The Notification Settings help you customize the kind of notifications you receive and how they should appear on your Mac, so you don't get distracted but still catch the important updates.

### LAUNCH NOTIFICATION SETTINGS:

1) Open System Settings:
   a) Click on the Apple menu located at the topmost part of your screen.
   b) Click on System Settings (or click on System Preferences, depending on which one your OS shows).
2) Go to Notifications: Within the sidebar, find Notifications. This is where you choose how notifications appear and from which applications.

## SETTING UP NOTIFICATIONS

Now that you are inside the Notifications window, let's look at configuring your settings to your liking:

1) Choose Applications: A list of applications that can show notifications will appear on the left-hand side. You can click on any one of these applications to configure its notification settings.
2) Notification Options: For each application, there are a number of settings you can modify:
    a) Allow Notifications: Allow or block notifications from that app.
    b) Alert Style: There are three different ways to present notifications:
        i) Banners: temporary notifications that appear along the top and automatically disappear.
        ii) Alerts: notifications that stick around until you take an action.
        iii) None: you won't see any notifications
    c) Notification Sounds: turn sounds for notifications from that app on or off.
    d) Badge App Icon: This will enable you to choose whether you have a red badge on the app icon showing notifications are unread.
    e) Show Previews: This option is where you can choose if you view notifications-Always, When Unlocked, or Never.
    f) Notification Grouping: This will bring up the choice of notification by application, automatic, or off.
3) iPhone Notifications

a) If you happen to use an iPhone, you can select to view iPhone notifications on your Mac. This is very helpful in keeping you connected without having to switch between devices.
b) Look for iPhone notifications in the Notifications settings and select the option to enable receiving alerts directly from your iPhone onto your Mac.

**Conclusion**

Notifications in macOS provide a quick way to stay organized and informed by managing your notifications, interacting with reminders, and showing you important information. You have learnt how to open it, interact with the notifications, and adjust its settings to improve productivity and enhance your experience on Macs.

# CHAPTER 7: FOCUS

This feature helps you in minimizing distractions and filtering notifications based on your current activity such as work, relax or focus on family. This feature helps you in staying focused on what is important at that moment.

## SETTING UP FOCUS

1) Open Focus Settings: Open the System Settings menu and select Focus from the sidebar.
2) Choose one of the available options to enable Focus Mode:
   a) Mac OS automatically provides you with some standard Focus modes, such as Do not Disturb, Personal, and Work; you can choose any of these or, if desired, create one.
   b) To activate a focus mode, click on your preferred option.
3) Customize Settings for Focus:
   a) You can tailor each Focus mode to allow notifications from certain apps or contacts. That is, while you may mute most notifications, you may want to allow critical alerts to come through from colleagues or family members.
   b) Set up Focus for:
      i) People: Allow notifications from selected contacts-for instance, important calls or messages.

ii) Apps: Enable notifications from selected apps you consider essential for your current activity.

iii) Calls: Select to allow calls during your Focus time.

iv) Time-sensitive notifications: Enable notifications which are marked as timesensitive to bypass the Focus setting.

4) Share Focus Across Devices: If you have several Apple device, then you can share your Focus status across devices. This includes the fact that when you turn on Focus on one device it will automatically turn on across devices for consistency.

5) Focus Filters: Apply Focus filters to block potentially distracting content from appearing within selected apps, such as Calendar and Messages. This helps you stay focused on the work at hand without additional distractions.

## ENABLING AND DISABLING FOCUS

1) Quick Access:

   a) To turn Focus on and off quickly, click the Control Center icon - a pair of toggle switches - in the menu bar.

   b) Click the Focus section to open a list of available Focus modes, which you can enable.

   c) When Focus is on, Messages will display a status to your contacts that notifications are

silenced. This way, others will know you are not available.
2) Set duration of Focus: Specify how long a Focus should be on, or turn it automatically on and off with the detection of location or time of day. Use this feature to keep up with the better balance in your work-life routine.

# CHAPTER 8: ACCESSING AND CUSTOMIZING YOUR WIDGETS

These are small applications that display information from your favorite apps right on your desktop or in Notification Center. You can make it your workspace by customizing them. They can display everything from the weather to Calendar events and reminders, among many other things. The following tutorial will explain how to add and remove widgets and rearrange them, including how to use widgets from your iPhone and third-party apps.

### STEP 1: OPENING THE WIDGET GALLERY

1) Enter the Notification Center: Clicking the date and time in the the screen's upper-right corner opens the Notification Center. If you have a trackpad, you can also swipe left from the trackpad's right edge with two fingers.
2) Click Edit Widgets: At the bottom of the Notification Center, click on Edit Widgets. The widget gallery opens-that is, the view where available widgets are shown, organized by app.

### ADDING WIDGETS

1) Widget Gallery Browse: You scroll through the apps and available widgets in the gallery, for which you will notice the variations in the sizes

and types depending on what's relevant in the functional knowledge of an app. Say, a weather app may allow small, medium, and large-sized widgets where it would show current conditions, forecasts, or alerts.
2) Add Widgets to Your Desktop:
   a) To add a widget, just click the widget size you want, and the widget will now appear inside your Notification Center. Drag and drop it onto the desktop if you prefer to get it outside of Notification Center.
   b) Sing in with the same Apple ID to both your iPhone and Mac, you easily can add iPhone widgets to your Mac Desktop. Moreover, this will let you view lots of useful information without a need to install the corresponding apps on your Mac. Just find the widgets you have set up already on your iPhone.
3) Drag and Drop: Once added, the widget can be dragged to where you want it to sit in either the desktop or Notification Center. Place it where you can easily view it and arrange the layout to suit your workflow.

## STEP 3: REMOVING WIDGETS

1) Enter Edit Mode: Click again at the bottom of Notification Center to return to the Edit Widgets menu.
2) Remove Widgets:
   a) Locate the widget in the Notification Center which you'll like to remove. Click the minus (-)

that shows up when you hover over it and the widget will disappear from your sight.

b) When removing a widget, you aren't uninstalling the app but just concealing the widget from your desktop or Notification Center. You can add it any time later by returning to the widget gallery.

### STEP 4: REARRANGING WIDGETS

1) Drag to Rearrange:
   a) You can even reorder the widgets within the Edit Widgets mode. Click and drag any of the widgets to a new position, either within the Notification Center or on your desktop.
   b) The widgets will snap into place so you can create a customized layout that prioritizes the information most important to you.

### ADDING THIRD-PARTY WIDGETS

macOS also supports third-party widgets, which you can download from the App Store. These widgets will extend your desktop experience by offering added functionality that might not be available through the default macOS applications.

1) App Store: Open the App Store either from your Dock or through a Spotlight search by hitting Command + Space and typing "App Store".
2) Search for Widgets: Search for widget applications using the App Store's search or by heading to the "Widgets" category, if present. The trendy widget apps can give you options on how to

manage your tasks, record fitness, or check more detailed weather.

3) Get the App Installed: Once you get your desired application offering widgets, click Get or Download to install it.

4) Add Third-Party Widgets: Now, to access Edit Widgets again, repeat the steps above. Now it will be in your widget gallery. Drag its widgets onto your desktop or to Notification Center like you would any other widget.

## CUSTOMIZING YOUR WIDGETS TIPS

1) Mix and Match Sizes: Sometimes these variably-sized widgets will be capable of displaying variable amounts of information. You might make a few small, a few medium-sized, and a few large widgets in order to experiment with which you like best.

2) Group Similar Widgets: If you have a group of multiple widgets that all relate to similar information-say, calendar and reminders-you might want to group them together in such a way that a cohesive panel of information is presented.

3) Updates: From time to time, check for updates of those applications providing you with widgets. Developers might provide additional features or widgets that may be helpful.

4) Accessibility Features: Check, where available, the settings of your widgets if you have specific accessibility needs for color scheme and size preferences.

# CHAPTER 9: ACCESSING CONTROL CENTER

The Control Center on your Mac puts frequently used settings in one easy-to-access place, right there in the upper right corner of the screen. That makes this even quicker because you will not have to navigate through several menus to adjust the settings you need. Instead, follow these steps:

## OPEN CONTROL CENTER

To do so, click the Control Center icon - it resembles two stacked switches - in the the menu bar's upper-right corner. This opens the Control Center panel, which includes most of your Mac's settings, arranged by category.

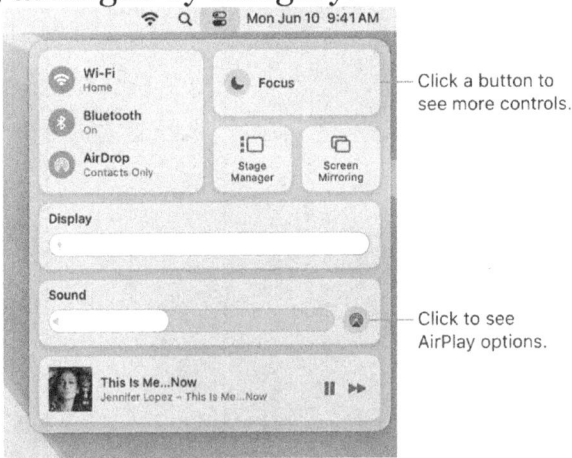

## KEY FEATURES OF CONTROL CENTER

Once opened, you will notice that Control Center has a number of key controls in sections including:

## BLUETOOTH AND WI-FI

1) Bluetooth:
   a) Bluetooth button: The Bluetooth button is used to enable or disable Bluetooth. Upon clicking the button, a drop-down menu appears displaying the connected devices and options available. You can connect or disconnect Bluetooth devices like headphones or mice here.
2) Wi-Fi:
   a) Click the Wi-Fi 🛜 button to display all the available networks. From this menu, you can switch your network, view your preferred connections, and enter further options that concern Wi-Fi.
   b) You can also click on Open Wi-Fi Settings to dive deeper into the network configurations.

## BRIGHTNESS AND VOLUME CONTROLS

1) Brightness: The Brightness slider controls screen brightness. This feature is handy when working with variable light conditions since you have a fast way to make changes without going through the display settings for changes.
2) Volume: The volume slider allows the adjustment of the audio output. Click the button for options to

adjust the system sound, and also controls for the connected audio devices.

## ADDITIONAL OPTIONS

1) Click for More Options: There are numerous buttons in the Control Center that have more options upon a click. By clicking the Sound icon, for example, it can allow you to select audio output devices; you could switch from internal speakers, headphones, or external speakers.
2) Back to Main View: In the case of exploring extra options in the Control Center, clicking the icon of the Control Center ⊙ once more will take you back to its main view. Such a design keeps things hassle-free and speedy.

## MASTERING YOUR DESKTOP WITH STAGE MANAGER

Along with the Control Center, another feature boosting your productivity by organizing your work area is Stage Manager.

### WHAT IS STAGE MANAGER?

Stage Manager means the feature which lets you to automate the setup of your opened apps and windows. It will help organize your desktop better to focus on what is important for you at one moment but easily switch between applications.

## HOW TO USE STAGE MANAGER

1) Enable Stage Manager:
   a) To enable Stage Manager, open Control Center and tap the Stage Manager icon-it looks like a window with two other overlaid apps on top of one another. Tap to turn Stage Manager on.
   b) You can also enable it from the System Settings > Desktop & Dock; it has a toggle for Stage Manager.
2) Window Arrangement:
   a) Once Stage Manager is activated, open windows will align to the left-hand side of your screen. It keeps all your currently active applications in easy view without cluttering up your desktop with too many opened windows.
   b) Click any application from this list to make it the focus window; the other windows will remain partially visible at the side for easy switching between them.
3) Workspaces Creation
   a) You can also create custom workspaces by grouping related apps together. For example, if you are working on a presentation and want to refer to a document you will have both apps in one workspace. To group apps together you simply drag and drop them into the Stage Manager area.

## MICROPHONE RECORDING INDICATOR

macOS includes a new built-in security feature that lets you know when your microphone is being used. It increases the sense of privacy and security by letting your view which applications are using your microphone.

## APPEARANCE

When an application accesses the microphone, an on-record indicator - a small dot or an icon - appears at the top of the Control Centre. This may be helpful to ensure unauthorized applications without your knowledge cannot access your microphone.

## INTERPRETING THE INDICATOR

An on-screen indicator is a good way to let you know when applications are currently using your microphone. If you happen to notice this 'on' indicator and happen not to be using an application that should have access to the microphone, it's probably a good idea to investigate further and know which application is making use of your microphone, with whatever action that warrants-quit the app, change permissions, etc.

## MANAGE WHICH APPS CAN ACCESS THE MICROPHONE

To control permissions for what applications get to use your microphone, navigate to System Settings > Privacy & Security > Microphone. Here you'll be able to turn access off on a per-app basis, giving you more granular control of precisely which apps can use your microphone.

## PINNING CONTROL CENTER FAVORITES

Control Center gives you fast access to settings of importance, and you could further your productivity by making sure your favorite things are pinned in the menu bar for even quicker ways of accessing them without having to always hover through the Control Center every time a setting change is needed.

1) Drag to Pin Favorites:
   a) To pin an item to the menu bar, click the icon on the upper right corner of your screen to open Control Center. Once open, locate your desired item to be pinned-example, Wi-Fi, Bluetooth, or Volume-and drag that icon into your menu bar at the top of your screen.
   b) When you release, it will pin the item into the menu bar for easy access.
2) Personalising Menu Bar:
   a) Adjust Menu Bar Items: In the System Settings, go to Control Center, where you can see all the items available in the Control Center.
   b) Choose "Show in Menu Bar" from the pop-up menu set beside each item. In this way, you will be able to enable different controls on and off to include only those that you will most likely need.

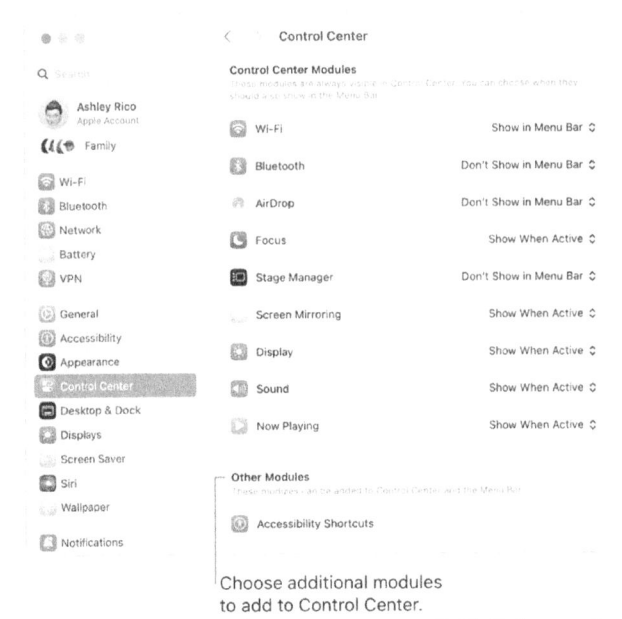

3) Preview Item Placement: Within the same settings menu, a preview of where the control will appear in the menu bar is available. This feature ensures that the controls can be organized in a manner that makes sense for your workflow.
4) Limitations: Some items cannot be added or removed from the Control Center or to the menu bar. By nature, they limit functionality and security by possessing accesses to system-critical controls.

## REMOVING ITEMS FROM MENU BAR

At one time or another, you may want to clean up your menu bar by removing things you either no longer use or need quick access to.

## Quick Removal Process With the Command Key

1) To remove an item quickly from the menu bar, hit the Command key ⌘ on your keyboard. While holding the key click and drag the desired item from off the menu bar.
2) When you let go of the mouse button, the item will disappear from the menu bar; this, however doesn't mean it has disappeared for good as it's still available in the Control Center.

# CHAPTER 10: WHAT IS SPOTLIGHT?

Spotlight 🔍 is an integrated search utility in macOS that allows you to find documents, images, applications, emails, and more on your Mac with ease. It's designed to make access to information fast without having to go through numerous folders or applications.

## OPENING SPOTLIGHT

Here's how to open Spotlight:

1) Using the Mouse: Click the Spotlight icon 🔍, a magnifying glass available inside the menu bar at your screen's top right.
2) Using the Keyboard:
   a) You can also use a keyboard shortcut for faster access. Press Command ⌘ + Space bar to open Spotlight. This is one of the fastest ways to initiate a search, and it's highly recommended if you want to be efficient.
   b) Spotlight Key F4: With some keyboards, you may have a dedicated Spotlight key F4. Clicking this will also activate the Spotlight search field.

## SEARCHING FOR ITEMS

Here are two ways of performing searches:

1) Type: Once you open the Spotlight 🔍, you can already begin typing what you are looking for. It will also show real-time results in Applications, Documents, Emails, Calendar Events, and Web Results as you type-on-the-go.
2) Live Text Feature: One of the useful things with Spotlight is that it allows you to search for text even within images. This is helpful, especially if what you want to find are screenshots of photos that have a written content. Note that this Live Text is available in some selected languages.

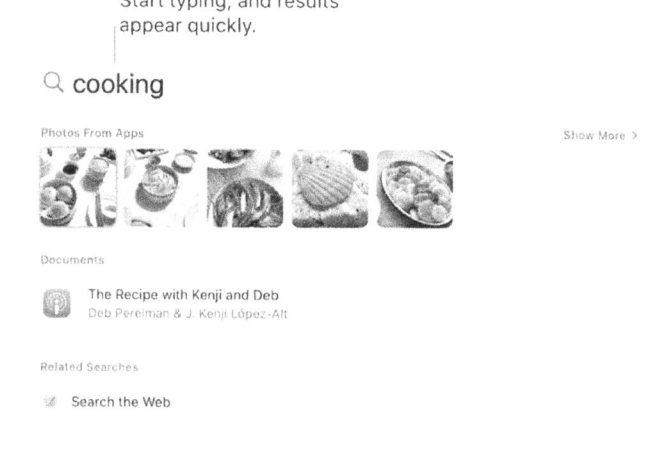

## OPENING APPLICATIONS

Here are two ways of opening applications using Spotlight:
1) Type App Name: When you need to open an application, you only need to type the name in the

85

search area of Spotlight. As an example, if you type "Safari", the Safari browser opens up.
2) Choose and Open: From the results of applications you've looked up, use Return to launch it instantly. You'll save your time rather than going through Applications or Launchpad yourself to open apps.

## PERFORMING QUICK ACTIONS

Here are some tips for performing Quick Actions with Spotlight:
1) Launch Quick Actions: Besides searching for files or apps, Spotlight can also perform quick actions. You can use it to run shortcuts, toggle settings like Do Not Disturb on or off, and to set times and alarms.
2) Search for Actions: Open Spotlight and start typing what you want to do. For instance, if you type "Turn on Do Not Disturb", you'll find that option available to turn it on right from your search results.
3) Examples of Quick Actions: You can trigger the following system commands with typing:
   a) "Turn on Do Not Disturb" to toggle notifications off.
   b) "Set alarm for 7 AM" to set an alarm.
   c) "Open System Preferences" to access immediate system settings.

## TIPS FOR EFFECTIVE USE OF SPOTLIGHT

Here're are some tips for optimizing your Spotlight searches:

1) Refine Your Searches: If your search results are too broad, you can narrow them down by including specific keywords or by using filters. You could also try typing "Documents" after your search term.
2) Use Natural Language: Spotlight is designed to understand natural language queries. For example, you could type "Show me pictures from last summer" to find relevant images easily.
3) Access Web Results: Spotlight can pull in results from the web, in addition to local files and applications, making it a very useful option for quick searching online.
4) Set up Spotlight Preferences: Configure Spotlight to your advantage, open the System Settings > Siri & Spotlight. You can select here which categories of indexing and displaying Spotlight shall be allowed to perform. This will make you focus more on relevant search results.

## CURRENCY, TEMPERATURE AND MEASUREMENT CONVERSIONS

< 1 cup

1 cup
0.5 pt

0.24 L
0.25 qt
23.66 cL
48 tsp

1. Spotlight can be quite handy in converting currencies without necessarily opening a separate converter application or going online to get conversions. Here's how you can use it effectively;
   a) Enter the Currency: Just type in the currency sign then the amount you want to be converted. Suppose you want to convert $100 into euros, you will type into "$100" in Spotlight and hit Return.
   b) View Converted Values: Spotlight displays a table with different values of the amount specified and it will be listed across various forms of currency. This feature will save you time from surfing the web by giving you the current conversion rates at your fingertips.
   c) Examples: Typing in "€50" will generate its equivalent in other currencies, such as dollars and yen.
1. Temperature Conversions:

a) Temperature Conversion: To convert temperatures, you can type a temperature value followed by its unit. For instance, to convert "32°C", you would type that directly into Spotlight.
b) Viewing Results: Press Return, and Spotlight will show the equivalent in Fahrenheit or other temperature units, thus easily changing over to a different measurement system.
2. Measurement Conversions:
a) Unit Conversions: You can also convert various measurements, including length, weight, and volume. In doing so, you will type the type of measurement you wish to convert, the amount, and the unit it is currently in. Example: If you type "10 miles to kilometers," it will return the number after it has been converted.
b) Examples: To convert "5 liters to gallons," you would type that into Spotlight, and it will return the amount in gallons.

## NARROWING YOUR SEARCH

Spotlight is configured to deliver a wide scope of results, but sometimes you want to refine the search term to be more specific: That is by narrowing, exclusion of specific folders, disks, and information types like emails or text messages. This becomes helpful in case you have volumes of data and want to situate yourself to particular areas.
1) Accessing Search Settings: To change these defaults open System Settings, then select

89

Spotlight. Once in this menu you want to select Search Results.
2) Choosing Categories: Once you're in this menu you'll be able to choose categories. For example if you don't like for Spotlight to go searching for Messages or Mail you'd de-select those options.
3) Advantages: The advantages here are that you can have much "cleaner" searches, seeing only what you are looking for.

# CHAPTER 11: USING SIRI SUGGESTIONS

Siri Suggestions is part of Spotlight and refines the search by offering contextually relevant information from all types of sources, including but not limited to:
1) What is Siri Suggestions? These are suggestions that draw information from sources such as Wikipedia, web searches, news articles, sports updates, weather forecasts, stocks, and movies.
2) How to Turn On Siri Suggestions: As soon as you start typing into Spotlight, related Siri Suggestions appear right in your search results for quick access to even more.
3) Search Information: For example type "weather," and Siri Suggestions may show you the current weather in your location, or various links about weather-related websites, news about weather events from around the world.
4) Constrain Siri Suggestions: If you would prefer Spotlight search only focus on things you have stored on your Mac, you can disable Siri Suggestions. For this, open the System Settings, click Spotlight and under Search Results uncheck the check box next to Siri Suggestions. This will ensure Spotlight will only show local files and applications in your search results.

## SETTING UP SIRI

1) Open System Settings: Click the Apple menu  at the top-left corner of the screen. Find the System Settings option there, but if working with a pre-2019 version of macOS, click the System Preferences instead. That would be used for altering several settings available on Mac.
2) Access Siri Settings: In the System Settings window, choose Siri & Spotlight from inside the sidebar. Here, all the Siri settings are housed.
3) Enable Siri: Find the **Enable Ask Siri** check box and click on it to enable Siri. This check box needs to be ticked. Now Siri will be ready for your orders.
4) Confirmation: After you turn Siri on, you might be prompted to confirm the action. Select the **Enable** Button to confirm the enabling.
5) Set up Siri preferences: Once you have turned Siri on, you can further edit several settings in order to personalize Siri as per your need:
    a) Language: Click the pop-up menu beside Language to choose the language you want Siri to use for responses and interactions. Siri is capable of using many languages. Please choose suitable for you.
    b) Voice: Select Voice from the Voice settings, and then select one of Siri's voices. You can choose between different accents and genders to your liking.
    c) Show in Menu Bar - If you want immediate access to Siri right from the menu bar, then select this check box to show Siri there. The option means that every time you want to use

Siri, you can click on the Siri icon at any moment in case you don't want to use a voice prompt.

6) Using the Dictation/Siri Key: If you're using a Magic Keyboard that has a Touch ID button, you can quickly access Siri using the F5 key 🎤 (which also functions as Dictation or Siri button) on your keyboard. This is a quick alternative to voice activation.

7) Setting Up Voice Activation: To activate Siri using your voice you may say "Hey Siri," or "Siri." To set this up:

　a) Select the System Settings again, then click on the Siri & Spotlight option.

　b) Find the option that says **Listen for** and click the associated pop-up menu. Here you can select either use Hey Siri or just Siri.

　c) Follow On-Screen Instructions: Once you have chosen to turn voice activation on, you will be taken through a set-up process, including perhaps repeating a few phrases to help Siri recognize your voice and fine-tune its accuracy.

## TURNING ON SIRI

1) Turn on Siri: After setting up Siri in the System settings, there are multiple ways to turn it on, making it very easy to access:
   a) Keyboard shortcut: Press the Command ⌘ and Space bar keys together. This is a fast shortcut to invoke the Siri:
   b) Menu Bar Icon: If you have set it to appear in the menu bar, you'll click on the Siri icon at the top right of your screen that looks like a little waveform or a little circle. This approach will be useful for those of you who like to see it to click it.
   c) Voice Activation: If you have enable the feature "Hey Siri," you can say it: "Hey Siri" to immediately access Siri. It is a good hands-free

option to multitask or when both of your hands are busy doing something else.
2) More About Siri: To learn more things Siri can do, say, "What can you do?" A list of the things Siri can do will appear. You can also go to the Apple website: Apple.com/siri for detailed information about Siri.
3) Type to Siri: If you don't wish to talk to Siri or you are in a surrounding that has high levels of noise, you can type your request.
   a) Activate Siri: Tap the Siri icon inside the menu bar. Alternatively, press hot keys Function (Fn) or Globe key ⊕ + S to turn on Siri.
   b) Typing Your Command: Once Siri is activated, you will encounter the Siri interface where you can type what you want to ask or order. This facility will be useful for those users who do not want to speak the command or those who have hearing impairments.
4) Show Captions: To hear and improve accessibility for what Siri is responding to, turn on the captions:
   a) Open System Settings from the Apple menu in the top-left corner of your screen. Note: In newer macOS versions, this is labelled as System Settings; on older versions, it's System Preferences.
   b) Select Siri & Spotlight in the sidebar.
   c) Enable Captions: Under the Siri Responses section, make sure there is a check next to "Always show Siri captions." This means every

95

response that Siri says will also be written on your screen for you to read.
5) Customize Siri Voice: In this section, learn how to change your Siri Voice settings:
   a) Open Siri Settings: Go back to System Settings and click on Siri & Spotlight.
   b) Siri Voice Selection: Locate the menu called Siri Voice. You will see voice variants, each different from others with regard to accent and gender. Tap the dropdown to listen to the different voices.
   c) Voice Options: Depending on what OS you are running, you may have a few voice options on your Mac. You can choose to have Siri to use another language, accent, or even a different gender. Choose your favorite and from that point forward, that is how Siri will sound.

## USING SIRI FOR EVERYDAY TASKS

Siri can assist you in almost everything, from the simplest to the most complex tasks. Following are some of the very helpful voice commands you may want to use:
1) Events Scheduling and Management:
   a) Calendar Management: You easily set up meetings or create reminders. You could say for example:
   Siri: "Create a new meeting for tomorrow at 2 PM."
   Siri: "Add a reminder to call Mom at 5 PM."

b) Checking Your Calendar: To check upcoming events, just ask:
  Siri: "Show me my calendar for this week."
2) Finding Information:
  a) General Queries: For anything you want to know, Siri can answer it. Example:
  Siri: "How high is Mount Whitney?"
  Siri: "What time is it in Paris?"
  b) Navigational Assistance: If you are wondering how to get somewhere, you can say:
  Siri: "How do I get home from here?" Siri will give you directions using Apple Maps.
3) Messaging and Calls:
  a) Messaging: Siri can save you from hassle in sending a text or message; it's as simple as:
  Siri: "Send a message to John saying I'll be late."
  b) Calling: You can also make calls with Siri:
  Siri: "Call Mom" or "FaceTime Sarah."

## MORE EXAMPLES OF SIRI COMMANDS

Here are more examples that will inspire your usage of Siri:
1. **General Questions:**
   Siri: "What's the weather like today?"
   Siri: "How tall is the Eiffel Tower?"
   Siri: "Tell me a joke."
2. **Productive Tasks:**
   Siri: "Create a new note."
   Siri: "Remind me to call Mom at 3 PM."
   Siri: "Show me my reminders."
3. **Navigation and Directions:**

Siri: "Directions to the nearest coffee shop."
Siri: "How do I get to the office?"
4. **Media Control:**
Siri: "Play some music."
Siri: "Pause video."
Siri: "What's the latest news?"
5. **Device Control:**
Siri: "Turn on Do Not Disturb."
Siri: "Dim the brightness to 50%."

## DRAWBACKS AND REMARKS

1. Internet Connectivity: Siri will only work on Mac if it has access to the internet. It is through the internet that Siri pulls information, answers questions, and executes commands.
2. Language and Regional Availability: Remember that Siri is not available in all languages and in all regions, and some of its features may be enabled only in certain parts of the world. So, if you try something and it doesn't work, check to see whether Siri supports this command in your native language.

## TROUBLESHOOTING SIRI

In case of any issues related to Siri:
1) Check Internet Connection: o Siri is an Internet-based assistant. Ensure that your Mac is connected to Wi-Fi.

2) Microphone Issues: Verify the microphone of your Mac. Do this by attempting other voice input features or applications.
3) Privacy Settings: If Siri does not respond to voice command, check your microphone privacy settings through System Settings under Privacy & Security.

# CHAPTER 12: USING ICLOUD WITH YOUR MACBOOK

iCloud allows your MacBook to interact seamlessly with other devices from Apple, keeping information such as documents, photos, and other data from your iPhone, iPad, or even Apple Watch in sync. Using the same Apple ID on all of them, iCloud will make sure you get any document, photo, and anything else at any given time so that you're always up to date and collaborate with your family and friends in the smoothest manner.

## SIGNING IN AND SETTING UP ICLOUD

If you didn't enable iCloud during the initial setup of your MacBook, you can still activate it anytime by following these steps:

1. Open System Settings: From your MacBook desktop, click on the Apple menu in the upper-left corner of the screen and choose System Settings.
2. Sign in with Apple Account: The System Settings sidebar will have the option saying Sign in with your Apple Account. Click on it and log in with your Apple ID and password if you haven't already.
3. Turn on iCloud: After logging in, you will have an iCloud option. Click on that and toggle on the features according to preference.

## CONFIGURING ICLOUD FEATURES

Some of the aspects in iCloud are immensely helpful, and you can opt to enable them or turn them off, depending on your needs. The following include:

1. iCloud Drive: You can enable iCloud Drive to store documents securely, and they will appear on all your devices. Also, any edit you do to a file on one of them instantly updates across all your other devices connected by your Apple ID.
2. Photos: These you will share with the iCloud Photos feature, enabling you to update photos and videos on either your iPhone, iPad, or MacBook and then view them on any other device linked with the same account. • Mail, Contacts, and Calendar: Keep your mail, contacts, and calendar events up to date across your devices. It's quite helpful if you really rely on e-mail, your contact list, and your calendar or reminders.
3. Messages iCloud: This is an option used to keep your messages updated across devices, so that conversations appear uniformly regardless of whether you are working in your MacBook, iPhone, or iPad.
4. Safari: In regards to Safari, turn on iCloud and all of your bookmarks, tabs, and browsing history will be in sync between your devices, serving to facilitate an easy switch between your various devices.
5. Find My Mac: This iCloud security feature enables you to locate your MacBook if it ever gets stolen or lost. You are given the opportunity to remotely lock or erase your device if it is necessary.

The items above are toggled on or off in the settings for iCloud. These give you options on which items you want to sync and access across multiple devices.

iCloud does not only allow you to sync across different devices but enables sharing where one can share folders, documents-even albums-with others. You might want to do this for working on projects with colleagues, organizing family photos, or sharing files with friends.

## USING YOUR MACBOOK WITH OTHER DEVICES

With Continuity, you can do the following with your MacBook and other Apple devices that are signed in with the same Apple ID and connected to the same Wi-Fi network. Continuity provides several useful- and creative-ways to move what you're doing from one gadget to another:
1. Handoff: You can begin using one gadget and continue what where you stopped on another. Examples: You can start an email or a document on the iPhone and then finish it on the MacBook using Handoff. To pick up on another device, simply click the Handoff icon in your MacBook's Dock or your iPhone/iPad app switcher.
2. iPhone as Webcam: Your MacBook comes with Continuity Camera, that enables the use of your iPhone as a high-quality webcam. To use it, initiate a video call from any application, such as FaceTime, from your MacBook while keeping the

iPhone in its proximity. The MacBook will instantly identify and pair with your iPhone's camera to bring you the benefits of higher resolution, along with Portrait mode and Centre Stage.

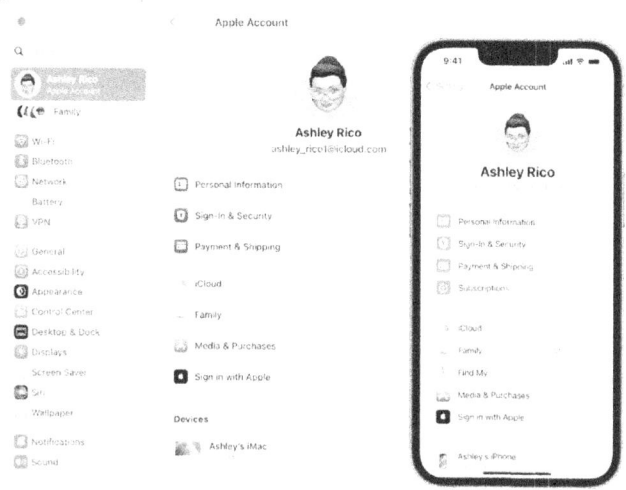

3. AirPlay: AirPlay enables streaming of what is on your MacBook to another compatible Apple device: such as a Mac, iPhone, and Apple TV. Great for either giving presentations, showing someone photos, or just watching videos on a bigger screen. You open the AirPlay menu on your MacBook and select which device you want to stream to.
4. Sidecar: This is going to let the use of your iPad as extended display for your MacBook. That's great for having extended workspace or multitasking without necessarily having another monitor. With the feature called Sidecar, go through your settings in the MacBook and under Display

settings, you will get a suggestion to add your nearby iPad as an extra display.
5. Universal Clipboard: Copy on one device, paste on another. For example, you copy some text or an image on your MacBook and directly paste it into a note or email on your iPhone or iPad, streamlining the workflow between all your devices.

## ACCESSING ICLOUD CONTENT ON YOUR MACBOOK

First set up iCloud on your MacBook, and have instant access to, and sync key content such as photos, files, notes, and passwords across all your Apple devices. Here's a close look at how iCloud keeps data connected and accessible on your MacBook and beyond.

### SETTING UP ICLOUD ON YOUR MACBOOK

1. Make Sure iCloud Is Turned On: If you didn't turn on iCloud during the setup process, open System Settings and click Apple ID; after that, sign in with your Apple ID. Further, select iCloud to turn it on.
2. Select iCloud Features: This will bring you to the options of all various things that are available to be selected, such as Photos, iCloud Drive, Notes, Safari, and Keychain. Turn on each feature you'd like to sync with iCloud; this information stores itself in the cloud and allows syncing automatically across your Apple devices.

3. Manage Your Storage: Apple provides everyone with 5 GB of free iCloud storage, but you can upgrade to iCloud+ to get more storage and other premium features. To check or upgrade your storage, go to System Settings > Apple ID > iCloud > Manage Storage. The iCloud+ plans house up to 2TB of storage alongside advanced features such as Private Relay and Hide My Email that help in privacy.

## ACCESSING YOUR ICLOUD CONTENT

After setting up iCloud on your MacBook, you'll have easy access to your content through several apps and locations that are specifically dedicated:

1. Photos: All the photos taken and videos recorded from your iPhone, iPad, or other Apple device will automatically appear in the Photos app on your MacBook. Open the Photos app to view, organize, and edit your images. Edits made on one device will be synced to all other devices, though you can elect to store full-resolution images on iCloud and lighter-resolution versions on your devices to save space.
2. iCloud Drive: You can access iCloud Drive directly from Finder on your MacBook for an easy way to save and manage files and folders in the cloud. You can drag files into iCloud Drive to access them from any other device signed into your Apple ID. For instance, a document saved to iCloud Drive on your MacBook can be opened and edited on your iPhone or iPad.

3. Notes: If you have Notes set to sync with iCloud, then your notes will stay up-to-date across all your devices. You write a note on your MacBook, and voil`, it's there in an instant on your iPhone, iPad, or other Apple devices.
4. Reminders and Calendars: Set up to-dos and appointments using Reminders and the Calendar app. You can have everything automatically set up across all your devices. Anything you edit or create on one device instantly updates everywhere.
5. iCloud Keychain: iCloud Keychain lets you save and share passwords, credit card numbers, and even Wi-Fi networks securely across the devices. It makes sure you always have the credentials you need for login on any device while keeping all this information encrypted for extra security. Keychain is controlled in System Settings > Passwords in your MacBook.
6. Safari Bookmarks and Tabs: These are also across devices with the use of Safari; just like your bookmarks, browsing history, and open tabs automatically keep themselves in sync. With Handoff or directly opening up open tabs through Safari, you can open a web page on your iOS device and pick up right where you stopped on your MacBook.

### ICLOUD+ BENEFITS

iCloud+ is an optional, paid upgrade to iCloud that affords you additional tools which could actually help with storage, privacy, and even collaboration. Here's what iCloud+ can do for you:

1. Expanded Storage: iCloud+ allows you to expand your storage beyond iCloud's free 5GB allowance. Plans start at 50GB and go up to 2TB, providing plenty of space for files, photos, backups, and more. You can share your iCloud+ storage plan with as many as five family members using Family Sharing.
2. Private Relay in iCloud: Private Relay refers to one of those privacy features related to encrypting your internet traffic during activity in Safari, hiding IP addresses to protect identity. It sends your internet activity through multiple servers, making it much harder for websites to track the online action.
3. HomeKit Secure Video: Recordings from smart security cameras can be stored securely by HomeKit with iCloud+, which doesn't take away your iCloud storage. This service includes end-to-end encryption, which means only you and the person you give access to will be able to see the recordings.
4. Custom Email Domain: This feature helps you connect your custom email domain to your iCloud account for sending and receiving emails with a personalized domain, such as "yourname@yourdomain.com." It makes communications sound more professional while still being hooked up with iCloud Mail.

## SYNCING ACROSS ALL DEVICES

Once you set up iCloud on your MacBook and all your other devices, Apple's ecosystem works in concert to automatically keep everything in sync. Use the same Apple ID on all of them, and voil-versa: files, photos, passwords, etc., appear everywhere. This further allows fluid and practically painless device switching, as iCloud works to put whatever you need at your fingertips-be it on your MacBook, iPhone, iPad, or Apple Watch.

## AUTOMATICALLY SAVE YOUR DESKTOP AND DOCUMENTS TO ICLOUD DRIVE WITH YOUR MACBOOK

Set up iCloud Drive to automatically save both your Desktop and Documents folders, and access your most-used files easier on all of your Apple devices. In such a case, any document that you would have saved on the Desktop or in the Documents folder of your

MacBook will immediately show up in the iCloud Drive, hence making it easy for you to access and update them from any other device of your choice. Here's how it works, and how to set it up:

## 1. HOW ICLOUD DRIVE WORKS WITH DESKTOP AND DOCUMENTS

By placing files on the Desktop or in the Documents folder on iCloud Drive, you create one spot for frequently accessed files. When you have iCloud Drive set up for the Desktop and Documents of a device:

1. The files that are stored in those folders are transferred to iCloud Drive and then immediately synced to other Apple devices that have the same Apple ID signed in to them.
2. You can access your files from anywhere—your iPhone, iPad, or another Mac—and even on non-Apple devices at iCloud.com or with the iCloud for Windows app.
3. Changes you make to files in these folders will be updated in real time across all your devices, so you're always working with the most current version.

This setup is ideal for working with document, spreadsheet, and image files that you're always updating, so that it's easy to switch between devices.

## 2. HOW TO CONFIGURE ICLOUD DRIVE FOR DESKTOP AND DOCUMENTS ON YOUR MACBOOK

To configure, follow these steps:
1. Launch System Settings: On your MacBook screen, from the top left, click the Apple menu and then select System Settings.
2. Click on Apple Account: On the sidebar, click on your Apple Account.
3. Enable iCloud Drive: Go to menu, select iCloud, then click on iCloud Drive.
4. Select Desktop and Documents Folders: You will notice you can choose the option to enable syncing for Desktop and Documents folders. This will save those particular folders in iCloud Drive. Confirm your selection in order to proceed.

## 3. ACCESSING YOUR FILES ACROSS DEVICES

Once you have enabled iCloud Drive for Desktop and Documents:
1. iPhone/iPad: After opening the Files app and going to iCloud Drive, you will notice that there are two folders: Desktop and Documents. Anything you save to those folders on your MacBook will appear here.
2. iCloud.com: For any web browser use your Apple ID to access your iCloud account at iCloud.com. Proceed to click on the iCloud Drive icon to access your files remotely.
3. From a Windows PC: From a Windows PC, begin by downloading & installing iCloud for Windows.

Subsequently, you will have access to the files stored in your iCloud Drive. Files on the MacBook that are stored in the Desktop and Documents folders can also appear in the iCloud Drive section on Windows.

### 4. BENEFITS OF STORING FILES WITHIN ICLOUD DRIVE

By having Desktop and Documents in iCloud Drive, you are able to:
1. Cloud Work: Your files are available on any device that has access to the internet. You can start working on a document on your MacBook and then pick up right where you stopped from your iPhone, or even your iPad.
2. Automatic Backups: The files in iCloud Drive will also create a backup in iCloud. It forms an extra layer of security. If your MacBook is lost, stolen, or suffers some kind of hardware failure, your files are safe in iCloud.
3. Partitioning Space: It automatically manages the storage of your MacBook by moving the older files to iCloud. The files are still accessible; however, because of inactive use, they remain stored in iCloud, which frees the local storage.

### 5. APPLY CHANGES AND DEVICE SYNCHRONIZATION

Any change or modification you make in the file on your MacBook will be saved automatically to iCloud

Drive and will be reflected instantly on another device.
3. Open and Edit Files: When you open a file on your MacBook, it downloads locally for immediate editing. When you save, it syncs back with iCloud Drive so any other connected devices have the latest version.
4. Offline Access: When opening files on your MacBook, they store locally so they will be accessible offline. If changes were made offline, they will sync back up with iCloud Drive when reconnected to the internet.

### 6. MANAGE STORAGE AND ICLOUD+

Each Apple ID is provided with 5 GB of free storage on iCloud, but iCloud+ provides the option to get more capacity along with additional premium features. You could upgrade to iCloud+ if you intend to keep a substantial amount of files or even larger types of files like videos or high-resolution images.

### 7. MORE ON USING ICLOUD DRIVE

1. Organize Files: Utilize folders in iCloud Drive to ensure a manageable Desktop and Documents across devices.
2. Manage Your MacBook Storage: With System Settings opening > select Apple ID > tap iCloud Drive, then toggle the button next to Optimize Mac Storage. This method will save space by storing only recently accessed files on your

MacBook while keeping others available in iCloud.
3. Sync Only What You Need: If you feel that iCloud Drive is syncing way too much data, then you can enable Desktop and Documents syncing to have certain files only on your MacBook.

# CHAPTER 13: STORING AND SHARING PHOTOS WITH ICLOUD ON YOUR MACBOOK

iCloud Photos allows you to access your entire photo and video collection, or share any amount of them, with ease across all your devices. Once iCloud Shared Photo Library is turned on, a space for shared media will be opened with other individuals, where each person can add photos, make edits, and even comment in real time. Here's how to set up both features and maximize their full potential.

## 1. SETTING UP ICLOUD PHOTOS ON YOUR MACBOOK

iCloud Photos keeps all your photos and videos in sync across your MacBook, iPhone, iPad, and other Apple devices connected with the same Apple ID. Whatever edits, deletions, or additions you make on one device will reflect on all your other devices.

1. Turn On System Settings: Open Your MacBook's System Settings.
2. iCloud Sign in: Go to your Apple Account inside the sidebar and click iCloud.
3. Turn On iCloud Photos: Under the iCloud settings, locate Photos, then turn it on. That'll cause iCloud to start syncing your photos and videos.

Once you have iCloud Photos turned on, it automatically launches the upload of your MacBook's Photos library media files to iCloud. You can then proceed to view or manage your photos from any device through its Photos app or via iCloud.com using non-Apple devices.

## 2. BENEFITS OF STORING YOUR PHOTOS IN ICLOUD PHOTOS

By enabling iCloud Photos on MacBook, it does the following:

1. Keeps Your Library Up-to-Date: It keeps all of your media and its edits up to date across each connected device. You will always have the most recent photo or video versions with you.
2. Provides a Safe Storage: iCloud gives safe backup for your photos and videos. If some problems appear with your MacBook, you will have the media in iCloud.
3. Optimizes Storage: macOS turns on the Optimize Mac Storage option by default in order to save storage. With this, iCloud contains full-resolution photos and videos, while your MacBook contains optimized versions to ensure at least local storage usage.

## 3. SETTING UP AND MANAGING ICLOUD SHARED PHOTO LIBRARY

With iCloud Photos, you can share a photo library with as many as five people. That's how you get to

share memories and moments with them. This will make all the members entitled to add, edit, and comment on the photos and videos so that it becomes shared space for each.
1. Open Photos: Open the Photos app on your MacBook.
2. Access Shared Library Settings: In the Photos app, go to Photos > Settings in the menu bar and then select the Shared Library tab.
3. Follow Onscreen Instructions: Tap Get Started to set up your Shared Library; follow the on-screen prompts that will pop up, inviting up to five others.

## INVITING PEOPLE AND SHARING PHOTOS

You can add members in it by inviting them through Messages or Email. After accepting the invitation, they have access and are able to add items to the shared library.

***Adding Content to the Shared Library:***
You are able to explicitly choose what photos and videos you'd like to share or use Smart Suggestions to add media based on your criteria such as:
1. Photos of specific people, using face detection,
2. Photos taken at specific places,
3. Pictures taken on a specific date or during some event.

This saves time and helps ensure that such important memories are not only located easily but also shared with people who matter.

## 4. HOW TO USE ICLOUD SHARED PHOTO LIBRARY FEATURES

In front of a Shared Library, everyone has equal rights to add or remove photos and videos, make edits, and add comments, which is highly collaborative. Unique features include:
1. Real-time Updates: Whatever one person edits gets updated in real time for everyone.
2. Edits and Comments: One can edit photos, including the usual ones like brightness and cropping, and even add comments for a collaborative feel.
3. Smart Curation: The AI in iCloud goes one step further to suggest the adding of photos, such as those with specific people or taken on specific dates.

## 5. MANAGING ICLOUD STORAGE

iCloud Storage provides 5 GB of storage with every Apple ID, and that fills up pretty fast if photos and videos are involved. To take on a larger library or add Shared Library members, you can: ICO
4. Upgrade to iCloud+: iCloud+ plans give extra storage and add premium features, which may be key in your case to store a serious amount of photos and video.
5. iCloud+: It is allowed to share an iCloud+ plan with family members to share storage space without each person purchasing one.

## 6. VIEWING PHOTOS ON ALL DEVICES

Once iCloud Photos has been activated:
2. With iPhone/iPad: Launch the Photos application and view your entire library.
3. Using a Windows PC: Set up iCloud for Windows; access your photos from File Explorer.
4. On iCloud.com: Sign in to iCloud.com using your Apple ID, then click on iCloud Photos.

## 7. PRIVACY AND SECURITY WITH ICLOUD PHOTOS

Apple puts a lot of importance on user privacy and especially when dealing with iCloud Photos. The feature is encrypted on an end-to-end basis; hence, only you and those whom you share the data with will access your photos and videos. Data encrypted in iCloud is stored by Apple, but Apple can never decrypt hence that data will remain private and secure.

# CHAPTER 14: ACCESS PURCHASES ANYWHERE WITH ICLOUD ON MACBOOK AND ENJOY

With iCloud, when you sign in to every one of your Apple devices using the same ID, it automatically syncs your purchases across your devices. Purchases through App Store, Apple TV app, Apple Books, and iTunes Store will show on all compatible devices that you sign in with your Apple ID. In particular, this is quite convenient for listening to music, watching movies, reading books, and using applications with your MacBook, iPhone, and iPad, Apple TV, and even Windows computers with iTunes.

## HOW TO ACCESS PURCHASES ON ALL DEVICES

1) Sign in using Apple ID: First, ensure you sign in to iCloud using one Apple ID for all devices; for your MacBook: open System Settings, then click your Apple Account in the sidebar, and under Apple ID, confirm that you're signed in.
2) To View Purchases:
    a) Music: Open Apple Music on your MacBook where you can view your purchased tracks, albums, and playlists.

b) Apple TV App: The Apple TV app can be opened to view any movies or shows that have been purchased or rented.
c) Books: With the Apple Books app, it's a snap to read any purchased eBooks or listen to any audiobooks.
d) Apps: Any apps you have bought on one device also appear on other devices under the Purchased section in the App Store.

With iCloud, you can start watching a movie on Apple TV, continue on your MacBook, and then continue from where you stopped on an iPad. iCloud keeps all of your purchases in the cloud.

## BENEFITS OF PURCHASING WITH ICLOUD SYNCING

1) Convenience-iCloud keeps purchases in sync for easy access to media across devices without needing to download the same files on multiple devices.
2) Consistent Experience: With Apple devices, you can continue right from where you stopped while using a book, movie, song, and others, on another device.
3) Storage Optimization: iCloud+ allows you to store your purchase in the cloud; hence, saving your device storage by only downloading the stuff whenever the need arises.

## USING FIND MY MAC TO PROTECT AND LOCATE YOUR MACBOOK

With Find My Mac, you can locate, lock, or even wipe your MacBook from iCloud if it happens to be lost or stolen. If Find My Mac is enabled, then use the Find My app on other Apple devices or iCloud.com to locate the last known location of your MacBook. It would have brought in peace of mind by availing options of remote access, which protect data in case your Mac is misplaced or compromised.

### SETTING UP FIND MY MAC

1) Turn on the Find My Mac Feature
   a) Open System Settings in your MacBook.
   b) Click on the Apple Account from the sidebar. Now tap **iCloud**.
   c) Under the iCloud settings, click Show All and toggle the **Find my Mac** to "on."
2) Using Find My Mac:
   a) Locate Your Device: If your MacBook was stolen, launch the Find My app on another device or go to iCloud.com, if you are on any other web browser. You could see it on a map here.
   b) Lock the Screen: Remotely locking your MacBook is possible. If you want to keep people from getting into your computer, and you can display a customized message when the screen is locked-on stuff like contact information.

c) Erase Data: If all else fails, then you can erase all data on your MacBook for preventing access to any personal or sensitive data.

## FIND MY MAC BENEFITS

1) Enhanced Security: Suppose someone tries to access your MacBook; you can easily lock it from a distance to keep your data protected.
2) Mind Find My Mac provides you with the reassurance of knowing that although your MacBook may be out of sight, you can take necessary steps in securing or recovering the same.
3) Remote Erase Knowing that you'll be able to erase all your data on Mac instantly in cases of an emergency adds to the security level, primarily concerning business or personal information.

## ICLOUD'S ROLE IN UNIFYING YOUR APPLE EXPERIENCE

iCloud allows your purchases and devices to connect seamlessly with one another into one continuous ecosystem that's easily accessible anywhere. With Find My Mac, you secure access to your MacBook without harming its safety. Both features, together, bring out the versatility and reliability of iCloud in simplifying your digital life while allowing you ways to protect it.

# CHAPTER 15: GIVING YOUR MACBOOK SOME PERSONALITY

That is because the personalization of your MacBook is going to introduce more comfort into the area you are sitting in and working with, along with a touch of uniqueness. This lesson will reveal how to change the wallpaper, attach widgets, and make your desktop look exactly the way you want it to be.

## HOW TO CHANGE THE WALLPAPER

Changes in wallpaper are among the easiest ways to give your MacBook a fresh look:

1) Open System Settings: Click the Apple menu  at the upper-left corner of your screen and then select System Settings, or System Preferences, based on your macOS version.
2) To get to Wallpaper: In the system settings sidebar, select Wallpaper. The window for setting wallpaper opens.
3) Select Your Wallpaper You will see some default wallpapers provided by Apple, or click **Desktop Pictures** to browse through a lot of categories.
4) To use a personal photo, click the Photos folder in the left sidebar to select an image from your library. Or, navigate to any specific folders where you store your images.

5) Select Wallpaper Style: Select the wallpaper to use with options like static or to dynamically change at set intervals such as every hour or day.
6) Match Your Screensaver: If you want everything to match, you could use the same picture for your wallpaper and your screensaver. You can also do this through Screensaver settings, which is still under System Settings.
7) Set the Wallpaper: Once you have your choice, click out of the settings. Your new wallpaper will take effect and give your desktop a new refreshed personal look.

## ADD WIDGETS TO YOUR DESKTOP

Widgets put key information and apps in view to make your desktop more functional and personalized to your needs. To do this, follow these steps:
1) Open Notification Center: Click the date or time in your screen's upper-right corner to open the

Notification Center. This is where you can access notifications and widgets.
2) Add Widgets: To add a widget to Notification Center, click Add Widget at the bottom. This opens the Widget Gallery where you can see several widgets that you can use on your desktop.

3) Widget Options: You shall be able to look at different types of widgets for addition, including:
   a) Photos: There you can put a photo widget inside, showcasing particular album photos or memories.
   b) Podcasts: A Podcast widget can let you know where you left off with your favorite shows and episodes.
   c) Calendar: Events widget will show what's next.
   d) Weather: You can add a weather widget that will show current conditions and forecasts.
4) Add and Rearrange Widgets:
   a) To add a widget, just click on it or drag it into Notification Center or to your desktop.

b) You can reorder widgets by dragging them onto your desktop or into the Notification Center.
5) Making Widget Size Changes: Most widgets come in a number of different sizes, including small, medium, and large. Frequently you'll be able to click and hold on a widget to show resize options where that makes sense for you and is supported on your screen.
6) Removing Widgets: To remove a widget from Notification Center if you no longer want it: Click and hold, then click Remove Widget or drag outside Notification Center.

## CUSTOMIZE YOUR DESKTOP ENVIRONMENT

Further customization of the desktop may involve:
1. Changing Desktop Icons: You can customize the icons of folders and applications on your desktop. Once you right click on the icon of your folder or application, select Get Info, then drag in an image onto the icon box to change it to the custom icon.
2. Organize Desktop Items: Create files and applications with folders. Right-click the desktop, then select New Folder. Identify the folder by file type stored inside it. Drag relevant items into that folder to keep the desktop clean.
3. Dock Settings: There are those things you cannot live without in your interface: The Dock. You can change its position - bottom, left or right side of the screen - or even its size from System Settings > Dock & Menu Bar. Consider enabling

Magnification to make icons larger when hovering over them.
4. Create Stacks: macOS offers you Stacks, which enable you to clean up the files on your desktop. Right-click the desktop and click "Use Stacks." This will then organize similar files (e.g., documents, images, etc.) into stacks, which allows for tidier desktops.

## CREATE A MEMOJI ON MACOS

One fun thing you can do on a Mac is creating a Memoji. As a matter of fact, that's pretty easy, and for many, that may be an exciting way to visually express themselves in messaging and other platforms. That means toning skin, to accessories, creating a Memoji image that is specifically you. Your Memoji adds an original touch and makes your experience of macOS really personal. It can be used as an account picture, in messaging apps, or even on the log-in screen. Here is an inclusive guide on how to create and customize your Memoji, which will make it yours.
1) Open System Settings: First, you need to get into the System Settings to start creating your Memoji:
   a) Opening the Apple Menu: Click on the Apple logo at your screen's top left.
   b) Choose System Settings: In the menu that drops, click on System Settings-or if you're using an older macOS version, it will say System Preferences.
2) In System Settings, go to **Users & Groups**: Now that you are inside the System Settings, scroll

down the sidebar of System Settings and click Users & Groups. This area deals with your user accounts and settings.

3) Choose a Profile Picture: You are now in the Users & Groups. Next to your login name, you will find your current profile picture. Tap this image to make changes to it.

4) Click Memoji: When the pop-up menu opens, you will have a number of options for changing your picture: Tap on **Memoji**. This will then open the Memoji creation interface.

5) Create Your Memoji: Since you have now opened the Memoji interface, you can go ahead to create your personalized emoji:

   a) Click the Plus Sign (+): To start creating a new Memoji, click on the plus sign. This will launch a set of customization options where you can design your Memoji.

   b) Personalize Your Memoji: You will be given a set of options to make a personalized Memoji. Some of the options you can personalize are:

   i) Skin Tone: You can choose from a wide range of skin tones that are pretty much close to your look.

   ii) Hairstyle: There, you will get an option to select different lengths and color of hairstyle. You can try different versions until you create one that accurately represents you.

- iii) Eyes: You can change the shape, color, and even add lashes to your Memoji's eyes if you want to give it a bit of personality.
- iv) Nostrils and Mouth: You get to choose nose shapes and colors of lips for finer details of your Memoji's face.
- v) Ears: Choose the ears' shapes and sizes that actually match yours.
- vi) Facial Hair: If you have some facial hair, then you could add mustaches, beards, and goatees, too, along with their colors.
- vii) Headwear: Go ahead to check out options for hats, eyeglasses, and more to complete your look.
- viii) Accessories: You can add unique accessories to your Memoji, such as earrings or sunglasses, to make it really different.

6) Preview Your Memoji: As you change your Memoji, a live preview of your creation will show up on screen. This is how you know exactly how every edit affects your Memoji in real time.

7) Save Your Memoji: Once you are satisfied with the design of your Memoji, tap Done. Your Memoji is saved now and ready for use.

8) Making your Memoji your Profile Picture: Once you've saved your Memoji, now you can set it as your profile picture:
   a) Select Your Memoji: You can see your newly created Memoji in the options of your picture in the Users & Groups settings again. Tap on it to set it as your profile image.

9) Using Your Memoji Across Apps: Now that your Memoji is ready, you can use it across apps like this:
   a) Apple Account Picture: Your Memoji will appear as an Apple account picture whenever you sign in using your Apple ID to any device or service.
   b) Contacts: You can set your Memoji in the "My Card" section in the Contacts app to identify yourself easily for others.
   c) Messaging and FaceTime: With Messages and FaceTime, use your Memoji to chat with friends and make video calls, putting a personal touch.
   d) Login Window: The Memoji will appear on the login screen to pictorially represent you at Mac startup.

# CHAPTER 16: WORKING WITH MULTIPLE SPACES ON MAC

To have maximum productivity in Mac, good management of your workspace is necessary. Though creating multiple desktops is certainly one of the most productive ways, managing them becomes more flexible through Mission Control, where created desktops are referred to as 'spaces'. It lets you arrange your opened windows and apps better, avoiding clutter and helping to focus on particular tasks. A step-by-step detail on how to create and manage spaces in a Mac is shown below.

### 1. UNDERSTANDING MISSION CONTROL

Mission Control is a powerful feature in macOS that gives you an overview of all your open windows, desktops, and full-screen apps. It lets you rapidly move between different spaces and manage your workspace more effectively.

### 2. ENTERING MISSION CONTROL

To create and manage spaces, first, you have to enter Mission Control. There are a couple of ways you could do this:
1) Using a Trackpad: Three-finger swipe up on your trackpad-a gesture that opens Mission Control, showing you all your open windows and spaces.
2) Use keyboard: Mission Control key, usually F3 or a key showing three rectangles, or press Control +

Up Arrow. This too takes you into Mission Control.

The Spaces bar

Create a space.

Access full screen or Split View apps.

## 3. CREATING A NEW SPACE

Spaces let you organize your opened applications and windows into separate desktops so that you can focus on specific tasks without visual clutter. When you are in Mission Control, you can simply create a new space:

1) The Spaces Bar: There, you will see the Spaces bar at the top of your screen displaying thumbnails of your desktops and full-screen apps currently running.
2) Click the Plus (+) Button: Locate the plus sign (+) on the right-hand side of the Spaces bar. Clicking that button creates a new space. You can create up to 16 spaces.
3) Finish Creating Your Spaces: When you click the plus button, a new space will pop up. This will reflect in the Spaces bar, where you can immediately have its thumbnail shown.

## 4. SWITCH BETWEEN SPACES

Once you have configured an area, it is relatively easy to switch between desktops in the following way. To switch between desktops:

1) Click the Space Thumbnail: In the Spaces bar, click on any one of the space thumbnails that you want to open. It will open up the view of that particular space.
2) Swipe between Spaces: Or if you are more into gestures, you can switch across your created spaces by swiping over the trackpad leftwards or rightwards using three fingers.

## 5. WORKING IN A SPACE

Working on a space, you would only see the opened windows and applications of that specific space; hence, your workflow would remain neat and clean and wouldn't get disorganized/disrupted by other opened applications.

## 6. CONFIGURING EACH SPACE

In order to make different spaces more personalized and thus easily distinguishable, you can assign different desktop pictures to each:

1) Set Wallpaper for All Desktops: Open System Settings: click the Apple menu, then select System Settings, or System Preferences. Under the Desktop & Screen Saver, you choose a wallpaper for all spaces.

2) Change Wallpaper for Individual Spaces: Go to Mission Control and switch to whatever space you want. Right-click on the desktop to select Change Desktop Background. Choose a different wallpaper for that space. Do this with each space if you want to establish a different look for them.

### 7. MANAGING SPACES

As your workflow changes, you may want to rearrange or remove spaces in any of these following manners:
1) Rearranging Spaces: In the Spaces bar, click and drag a space's thumbnail left or right to rearrange the order of your spaces.
2) Removing a Space: To delete a space, enter Mission Control, hover over the space you want to remove, and click the "X" that appears in the corner of the space thumbnail. This will delete that space.

## DIFFERENT WAYS TO SWITCH BETWEEN SPACES ON MAC

Switching between spaces on your Mac can make a big difference in how much work you get done, just by organizing your workspace. From Trackpad gestures to keyboard shortcuts and the Touch Bar, macOS offers a few ways to move around among multiple desktops with ease. These include the following:

## 1. TRACKPAD GESTURES

Your trackpad is multi-touch enabled, and this allows you to have some pretty natural gestures to move between spaces with ease. Here's how you do this:

1) Three-finger swipe-on most MacBooks and other trackpads, swiping left or right with three fingers will take you between spaces. This saves time because you won't be interrupting whatever you may be doing.

2) Four-Finger Swipe: Depending on your trackpad settings, you might also use a four-finger swipe to do this. This can be set in System Settings > Trackpad under the More Gestures tab. Use whichever is most comfortable to you.

## 2. USING A MAGIC MOUSE

Using a Magic Mouse is quite straightforward. You can easily move from one space to the other by performing simple gestures. These are:

1) Swipe with Two Fingers: Place two fingers on the Magic Mouse and swipe left or right to switch between spaces. This gesture emulates using a trackpad-swipe gesture and will smoothly transition between your organized workspace.

## 3. USING KEYBOARD SHORTCUTS

For those who are accustomed to-or merely partial to-keyboard control, macOS makes it quite easy to use keyboard shortcuts to navigate through spaces:

1) Control Key with Arrow Keys: Press and hold the Control key, then tap the Right Arrow key to head to the space on the right or the Left Arrow key to go to the space on the left. This is quite fast and efficient, especially when switching between spaces very often.

### 4. USING MISSION CONTROL

Mission Control provides an overview of all your opened windows and open spaces for easy navigation and management of your workspace. Following are some of the ways to manage the Mission Control application:

1) Entering Mission Control: To enter Mission Control, you have three options: you may swipe upwards with three fingers on your trackpad, hit the Mission Control key typically the F3, or click Control and the Up Arrow. After selecting Space via the Spaces Bar,
2) Once in Mission Control, move your cursor to the top edge of the screen, where you will see the Spaces bar. On the Spaces bar, you will see a thumbnail of all the active open spaces.
3) Click the thumbnail of the space you would like to switch to. This is pretty useful if you have many open spaces and you want an overview of them before selecting one.

## 5. USING THE TOUCH BAR (FOR MACBOOK PRO MODELS)

If you have a MacBook Pro with a Touch Bar, you can easily switch spaces using it:

1) Enable with Touch Bar: To be able to see your spaces in the Touch Bar, you first have to enable that option. Open System Settings and go to **Keyboard** and then select the option to show spaces in the Touch Bar.
2) Tap to Switch Spaces: When the feature is activated, the Touch Bar will show you thumbnails of your active spaces. Tap on the space you want to open and immediately get switched into that desktop.

## TIPS FOR SPACES USE

1) Keep it tidy: If you create so many spaces too frequently, then regular clean-up of those you don't use may keep the Spaces bar from cluttering and slow navigation.
2) Direct attention to specific tasks: Applications concerning one form of work must be opened within the same space. For instance, all design applications may be on one space, and all research ones on another. This simply said, will reduce the number of times you have to switch between spaces.
3) Employ full-screen apps: In case you are working with full-screen applications like Safari or your design program, they will automatically create a

new space. This too can be leveraged for minimizing distractions.
4) Practice Gesture Control: Familiarize yourself with the gestures for switching spacing in lightning speed with your trackpad or Magic Mouse. With more practice, it surely will become second nature.
5) Customize Settings: Go to System Settings > Trackpad or Mouse, where you can adjust the sensitivity to your liking and enable or disable certain gestures.

That's part of the beauty of Spaces in macOS: you can dictate that specific apps will always open in certain spaces. This cleans up your workflow by decreasing distractions, and everything is exactly where you expect it each time. Let's go through the process step-by-step, covering each of the options available as you assign apps to spaces.

## HOW TO ASSIGN APPS TO SPACES ON MACOS STEP BY STEP

1) Open the App you want:
   a) The first thing you do is fire up an application you would like to attach to a particular space. An already opened application will be shown in the Dock.
   b) Locate its icon in the Dock. Control-click or right-click the app icon to open a menu list of options.
2) Click 'Options' then Choose an 'Assign To' Option: Hover over Options in the menu, and you will see

a number of options appear under Assign To. You have a few configuration options; here each is a different way to handle managing your application's location across spaces:

### ASSIGN TO OPTIONS:

1) All Desktops:
   a) All Desktops: It selects all the desktops and opens the application in every space that you have created. This is useful for applications that might be needed whatever the workspace context, such as communication applications like Messages or Slack and productivity applications like Calendar or Reminders.
   b) Opening the app in every space means its windows will follow you when switching between spaces, hence reducing the need to toggle between different screens.
2) This Desktop:
   a) Choose This Desktop keeps the application in the current space. Every time you open this application, by default it opens in this space alone.
   b) The option is pretty handy for task-related applications. You might want to keep design tools like Adobe Photoshop or Illustrator in one space while your coding tools are in Xcode or Terminal in another space.
   c) If you have selected full-screen for the app to open, then it opens a space unto its own in that desktop that you choose.

3) Desktop on Display [Number]:
   a) If you are working with more than one display then macOS allows you to assign an app to a particular display for a specific desktop. It will be labeled Desktop on Display [number] in the menu where each external monitor or built-in display has been assigned a different number.
   b) This option is useful in dual-display setups where you'd want a video call app, say, always on Display 1, and a presentation tool open on Display 2, each confined to their spaces for easy access and maximum use of screen estate.
4) NONE:
   a) Choosing None clears any space assignment for the application. That is to say, the application will open in whatever space you happen to be using. If you don't need to have a particular workspace configuration, this keeps things flexible and allows the application to open wherever you may happen to be.
   b) NONE - Apps are best designated for applications that are occasional-use only. It would be irrelevant to give a specific space designation for system utilities or to productivity applications that see very little use.

## MORE TIPS TO MANAGE APP BEHAVIOR IN SPACES

1) Automatic Switching to a Space with an Already Open Window:

a) By default, if you switch into an application, macOS automatically switches to the space where an already open window of that app resides. This default behavior is pretty useful when you want to avoid opening a duplicate window and jumping between spaces.
b) For example, if you have a document open in **TextEdit** in Desktop 3 and then click to create a new TextEdit window in Desktop 2, macOS will whisk you back to Desktop 3 where the open windows are already showing.

2) System Settings: How to Customize This Behavior
   a) To change this behavior, open the Apple menu and click System Settings. Click **Desktop & Dock** 🖥 in the sidebar and scroll down to Mission Control. Go to the option for Moving to a Space where the application's windows are open when you want to switch apps.
   b) You can turn this on and off as you please. Off continues to open a new app window in your current space, which is great if you like to keep workspaces strictly separated.

## PRACTICAL APPLICATIONS OF ASSIGNING APPS TO SPACES

App assignment to spaces can be used to extend productivity by offering a clean, separated space for related tasks. This is for focused and segregated task separation in work environments. The given

scenarios are quintessence use of the feature for the following:
1) Focus and Task Segregation: Assign the productivity apps-Word and Excel-to one space, creative apps like Photoshop and Illustrator to another, and comms apps like Slack or Zoom to a third. In this way, you will be able to enter the appropriate space depending on the task at hand, helping you with focused work without clutter in the mind.
2) Multiple Display Setups: Assign certain apps to specific desktops on certain displays, especially for users with dual or triple monitor setups. Examples might include reserving Display 1 for project management apps, Display 2 for coding or content creation, and Display 3 for web browsing or research.
3) Entertainment and Break Spaces: Or make a space dedicated only to non-work applications: music, video sharing sites, news sites, whatever. Then, when you feel like taking a break, switch to the 'entertainment' space without fiddling with your main workspaces.

## MOVING AN APPLICATION WINDOW FROM ONE SPACE TO THE OTHER

Working with more than one space open in macOS, each can be dedicated to certain tasks or workflows, the capability of moving windows across is priceless to adapt to the priorities that change. There are two basic ways to transfer app windows from one space to the other.

## METHOD 1: DRAGGING TO THE EDGE OF THE SCREEN

4) Select the App Window to Move: Click and hold down on the title bar of the window of the application you want to move.
2. Drag to Screen's Edge: With the window of the application held, drag it to either the screen's left or right edge. After a short while, macOS will automatically bring the window into the next adjacent space in that direction.
3. Continue to Adjacent Spaces:
    a) To take it further, pull the window to the screen's edge until you reach where you would want to create space.
    b) Practical Use Case: This is helpful if you have, for example, moved windows between spaces in succession-for instance, taking a project management application to a workspace closer to your current space.

## METHOD 2: MOVING WINDOWS VIA MISSION CONTROL

1) Access Mission Control:
    a) Three-finger swipe up on trackpad, using Mission Control key-F3-on keyboard, or Control-Up Arrow shortcut.

   b) In Mission Control, there is a Spaces bar right across your screen's top with thumbnails of each space.
2) Drag and Drop the Window: Locate the window of the app you want to move in the active space section and drag it to any other space in the Spaces bar.
3) Put Windows in Split View:
   a) If you drag the app window onto the full-screen app thumbnail, macOS will automatically place the two apps in Split View within that space so that you can work with them side by side.
   b) Practical Use Case: You will want to use this approach when you're moving around between several window reorganizations or navigating between distantly placed spaces. This is because Mission Control gives an overview of all your spaces and all open windows.

## DELETING A SPACE

If you no longer need a certain space, you are able to easily delete it, which might help clean up your workspace from clutter. Performing the deletion of a space on macOS is pretty easy, though for that you need access to Mission Control.
1) Enter Mission Control: To open Mission Control, use any of the following: three-finger swipe upwards on the trackpad, Mission Control key (F3), or Control-Up Arrow.

2) Go to the Spaces Bar: At the top of the screen, thumbnails of each open space appear; this is called the Spaces bar.
3) Select the Space You'd like deleted: Pass the cursor over the space you'd like to have deleted, and an "X" icon will appear in the corner of that thumbnail.
4) Delete the Space: Once you clicked the "X" to delete the space: the open windows of that space automatically will move into another space. Hence you will never lose access to any of your files or applications.
5) Real-world Application: You would want to delete a space when you want to go back to having fewer workspaces once for example you have finished a project so you don't have a cluttered interface.

## EXITING FULL-SCREEN OR SPLIT VIEW IN SPACES

Sometimes you want to remove an application from full screen or Split View, but not the space itself. To leave full screen or Split View places that app back into a windowed mode in the current space.
1) Enter Mission Control: Open Mission Control using one of the methods outlined above.
2) Hover Over the Full-Screen or Split View Thumbnail: Find the full screen or Split View thumbnail for the application in the Spaces bar.
3) Exit the Full-Screen or Split View Mode:

a) Move the cursor onto the image in the thumbnail until you see an Exit button ⤢, click that. This will shut the full-screen mode down and send the app back into its normal window in the space.
b) Practical Use Case: this is useful if you want to quickly switch an app to windowed mode to multitask with other apps while still having them stay in the same space.

## BEST PRACTICES FOR MANAGING SPACES

You can work much more productively with macOS if you effectively use and handle the spaces. Here are a few tips to help you keep your workspace organized and productive:

1) Plan Your Spaces by Task or Project: Create different spaces for different tasks that you do, like work, personal, study, etc., and that way you can keep yourself focused on your current work. Create related apps in every space so you can easily shift between a really focused environment without being distracted by unrelated apps or files.
2) Limit the Number of Spaces: While macOS supports up to 16 spaces, too many of them will make it a nightmare to remember where everything is. Stick to not having more than you actually need at any given time.

3) Use Split View for Efficiency: Split View is ideal when multitasking on your computer, opening two items side by side. For example, open a reference document and a writing app together to streamline your work.
4) Customize Your Spaces with Unique Wallpapers: Assign different wallpapers to spaces, making them visually different for letting your memory know which space is used for what.
5) Mission Control Shortcuts to Quickly Access: Master Mission Control shortcuts, such as using Control-Up Arrow and trackpad gestures, to see how much of a time-saver it is to access and organize your spaces.

# CHAPTER 17: SYSTEM SETTINGS

To tailor your MacBook experience, System Settings provides a central hub for modifying a wide range of options, from basic screen adjustments to accessibility features. Here's how you can navigate and modify these settings to personalize your MacBook:

## ACCESSING SYSTEM SETTINGS

1) Open System Settings: Start by clicking the System Settings icon  directly in the Dock, or select it through the Apple menu at the top left of your screen, choosing  (Apple Menu) > System Settings.

2) Navigating the Sidebar: The left sidebar contains a list of settings categories. Scroll down if needed to view all available options, and click on the specific setting you wish to change.

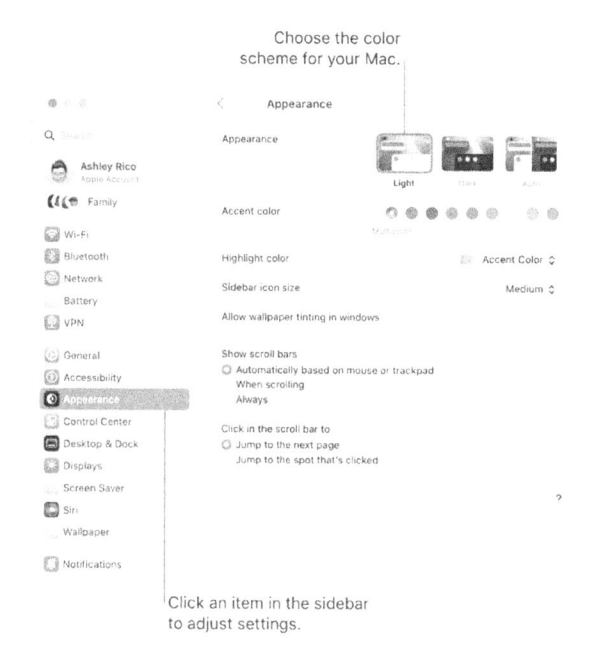

## MODIFYING KEY SETTINGS

1) Screen Saver and Wallpaper: Customize your display by choosing a screen saver, which you can also set as your wallpaper. This provides a cohesive visual experience and adds a personal touch to your desktop.
2) Notifications: Manage app alerts and notifications, tailoring which apps can send you alerts, when you'll receive them, and how they'll appear on your screen.
3) Display Settings: Adjust key display aspects, including brightness, color settings, and screen resolution, to enhance visibility and comfort based on your usage.

149

4) Accessibility: Fine-tune settings to make your MacBook easier to use. Adjust features for vision, hearing, motor functions, and more to support a more inclusive experience.

## KEEPING SOFTWARE UP TO DATE

In System Settings, you can also check for and install the latest macOS updates. Keeping your system updated ensures you have the newest features and important security patches.

## FINDING SPECIFIC SETTINGS QUICKLY

If you're looking for a specific option but aren't sure where it's located, use the search bar at the top of the System Settings window. As you type, relevant settings will appear in the sidebar, allowing you to navigate directly to the option you need.

Each adjustment can significantly enhance your interaction with your MacBook, making System Settings a versatile tool to ensure your Mac is optimized to suit your preferences.

## LOCKING YOUR SCREEN

1) Setting Lock Screen Preferences: To secure your MacBook when you're not actively using it, adjust the Lock Screen settings to automatically off the screen or activate a screen saver when a set period of inactivity elapses.

2) Enabling Password Requirements: For an additional layer of security, you can require a

password to open your screen when you get back. To set up these options:
a) Open System Settings.
b) Select Lock Screen from the sidebar.
c) From here, configure the idle time before the screen locks and activate the password requirement option to secure access.

## CHOOSING AND CUSTOMIZING A SCREEN SAVER

1) Selecting a Screen Saver: Pick a screen saver which matches your style or mood by going to Screen Saver within System Settings. Apple offers a variety of screen savers, including Landscape, Cityscape, Earth, Underwater, and Shuffle Aerials themes.
2) Setting the Screen Saver as Wallpaper: If you'd like your screen saver to serve as your wallpaper, enable this option by turning on Show as Wallpaper in the Screen Saver settings. This transforms your display into a visually dynamic experience, even when you're not using your MacBook.

## CUSTOMIZING CONTROL CENTER AND MENU BAR

1) Choosing Control Center and Menu Bar Options: The Control Center on macOS gives you quick access to essential settings and frequently used tools. You can personalize the Control Center to fit

your usage by selecting which settings appear there, and even add shortcuts to the menu bar for easy access.
2) Adjusting Preferences: To customize:
   1. Go to System Settings.
   2. Click Control Center in the sidebar.
   3. From here, select the settings you want to include in the Control Center, and toggle the option to show them in the menu bar if desired.

## UPDATING MACOS

1) Accessing Software Update: Regular updates ensure your MacBook has the latest features, performance improvements, and security fixes. To check for updates:
   a) Open System Settings, then choose General from the sidebar.
   b) Click Software Update to see if a new version of macOS is available.
2) Enabling Automatic Updates: If you prefer automatic updates:
   a) In Software Update, toggle the "Automatically keep my Mac up to date" option.
   b) You can further specify whether you want macOS to automatically download updates, install system files, or include app updates from the App Store. This setup is highly beneficial to keep your MacBook secure and up-to-date with minimal effort.

## ICLOUD AND FAMILY SHARING SETTINGS

1) Signing into iCloud: iCloud syncs your files, photos, and app data across all your Apple devices, enhancing accessibility and convenience. To set up iCloud on your MacBook:
   a) Open System Settings, click on your Apple ID at the sidebar's top, then select iCloud.
   b) Sign in using your Apple ID if you haven't already. Once signed in, you'll see a list of apps and services that can use iCloud.
2) Managing iCloud Storage and App Access:
   a) In the iCloud menu, review which apps and services are using iCloud storage and toggle off those you don't need synced.
   b) To manage your storage space, click Manage to view your iCloud usage and select options like Upgrade Storage if you need more space.
3) Managing Family Sharing: With Family Sharing, you can share your purchases, iCloud storage, and Apple subscriptions with family members.
   a) In System Settings, under Apple ID, select Family Sharing.
   b) Follow the prompts to invite family members using their Apple IDs. Once they accept, you can share iCloud storage, Apple Music, Apple TV+, and other services with them.
   c) You can also set up parental controls, manage Screen Time for each family member, even creating Apple ID for your child.

## ADJUSTING TRUE TONE FOR AMBIENT LIGHT

1) Understanding True Tone: True Tone technology automatically adjusts the color temperature of your MacBook's display based on the ambient lighting around you. This feature reduces the blue light from your screen in low-light environments and provides a more natural, eye-friendly viewing experience by adapting display warmth and coolness.
2) Turning True Tone On or Off:
   a) Open System Settings, then choose Displays from the sidebar.
   b) In the Displays section, toggle True Tone to enable or disable the feature based on your preference.
   c) When enabled, True Tone will seamlessly adjust display colors according to changes in ambient light, giving you the flexibility to work comfortably in different lighting conditions.

## SETTING UP DYNAMIC DESKTOP FOR TIME-BASED VISUAL CHANGES

1) Using Dynamic Desktop: Dynamic Desktop is a feature that changes your wallpaper automatically to match the time of day, offering a visually engaging experience as the screen brightness and color tones adjust subtly throughout the day.
2) Selecting a Dynamic Desktop Wallpaper:

a) Open System Settings, then choose Wallpaper from the sidebar.
   b) Scroll through the available wallpapers and select a Dynamic Desktop option, such as a landscape or cityscape that will change from dawn to dusk.
3) Enabling Time-Based Changes:
   a) For the wallpaper to change automatically according to the time of day, Location Services should be turned on.
   b) If you've disabled Location Services, Dynamic Desktop will instead rely on the Date & Time settings, adjusting based on the time zone specified in your system.
   c) To enable Location Services, go to System Settings. Then click on **Privacy & Security** and choose **Location Services** and toggle it on.

## CHANGING HOW ITEMS APPEAR ON THE SCREEN

1) Adjusting Display Resolution:
   a) To make items appear larger or smaller, you can modify the display resolution, which affects the size of everything on your screen, including windows, icons, and text.
   b) Go to System Settings and pick Displays from the sidebar.
   c) In the Displays settings, you'll see a section for Resolution. Choose Scaled to reveal available resolution options. Selecting a lower resolution

makes items larger, while higher resolutions reduce item size for a sharper display.
2) Increasing Text and Icon Size:
   a) For better readability, you can increase the size of icons and text without adjusting the full display resolution.
   b) Go to System Settings and pick Accessibility > Display. Here, you can make text larger and increase icon size, improving ease of viewing without impacting the overall resolution.
3) Making the Pointer Easier to See:
   a) You can adjust the appearance and visibility of the pointer to make it easier to find.
   b) In System Settings, go to Accessibility > Display > Pointer. Adjust the pointer's size and color here. You can also enable the "Shake mouse pointer to locate" feature, which momentarily enlarges the pointer when you shake your mouse, making it easy to find on a cluttered screen.

## ENABLING DARK MODE TO STAY FOCUSED

1) Activating Dark Mode:
   a) Dark Mode provides a dark color scheme for your desktop, menu bar, Dock, and system apps, improving focus by making content stand out and reducing glare.
   b) To enable Dark Mode, go to System Settings and pick Appearance from the sidebar.

c) In Appearance, choose Dark to activate Dark Mode across your MacBook.
2) Optimizing for Dark Environments:
   a) Dark Mode is especially useful for low-light environments, reducing eye strain and making it easier to view content without a bright screen.
   b) When Dark Mode is enabled, system apps like Mail, Calendar, Contacts, and Messages display white text on a black background, which is easier to read in dark settings.
3) Benefits for Professional Image Editing:
   a) Dark Mode is particularly beneficial for professionals working with images or videos, as it makes colors and fine details stand out against the darker background, offering better contrast.
   b) Designers, photographers, and other creative professionals can maintain focus on their work, as Dark Mode allows them to better view colors and details with minimal distraction.

## SETTING UP NIGHT SHIFT FOR WARMER COLORS

1) Understanding Night Shift: Night Shift reduces blue light by adjusting your screen to display warmer colors, which is beneficial at night or in low-light conditions. Research shows that blue light can interfere with sleep patterns, making it harder to wind down at the end of the day. Night

Shift is designed to help you reduce exposure to blue light, encouraging a more restful evening.
2) Activating Night Shift:
   a) Open System Settings and click Displays in the sidebar. At the bottom of the Displays settings, click Night Shift to access its options.
   b) You have a few options for activating Night Shift:
      i) Manual Activation: Turn Night Shift on or off manually as needed.
      ii) Scheduled Activation: Schedule Night Shift to automatically switch on and off at times that suit your routine. You can specify a custom schedule or set it to activate automatically from sunset to sunrise, using your time zone to determine activation.
3) Adjusting the Color Temperature:
   a) In the Night Shift settings, you'll see a Color Temperature slider. This slider allows you to fine-tune the warmth of your screen colors when Night Shift is active.
   b) Dragging the slider towards Less Warm will keep more blue light, while moving it towards More Warm will reduce blue light further, creating a warmer screen appearance.
   c) Test different levels of warmth to see which setting is most comfortable, especially in low-light or night time environments.

## CONNECTING AN EXTERNAL DISPLAY

1) Determining Display Support:

a) Before connecting an external display, projector, or HDTV, check how many external displays your MacBook supports. This can vary depending on the specific model and hardware capabilities.
b) To check this, go to System Settings and pick Help > MacBook Specifications from the menu. In the specifications, locate the Video Support section, where the number of supported external displays will be listed. You may need to scroll to find this section.

2) Connecting Your External Display:
   a) Your MacBook should have one or more Thunderbolt or HDMI ports for connecting external displays. Ensure you have the appropriate cable and, if necessary, an adapter.
   b) Connect the external display to the port on your MacBook. Your MacBook should automatically detect the new display and mirror or extend your screen based on your existing settings.

3) Configuring Display Settings:
   a) Once connected, Open System Settings, then choose Displays to configure how your MacBook interacts with the external display.
   b) You can choose to Mirror Displays if you want the same content to appear on both screens, or Use as Separate Display to extend your desktop and have additional screen space for multitasking.
   c) In the Displays settings, you'll also have the choice of setting the resolution and orientation

for each connected display, allowing you to customize each one according to your needs.

# CHAPTER 18: ACCESSIBILITY SETTINGS

Your MacBook has Accessibility settings that make using the computer easier for anyone, and especially so if someone has vision, hearing, or motor impairments. These options all range from larger text, distinguishing colors, making the mouse and keyboard more accessible, right to allowing you to control your MacBook with your voice. These will be covered in sections, entitled Vision, Hearing, Mobility, Speech, and General. Here's a full description of what each section comprises and what kind of customizations you can find under each heading.

## OPENING ACCESSIBILITY SETTINGS

To open the Accessibility settings:
1) Open the Apple menu and select System Settings.
2) Click Accessibility in the sidebar to see that the settings are organized by category: Vision, Hearing, Mobility, Speech, and General. Each of these categories provides a suite of tools designed for different accessibility needs.

## ACCESSIBILITY CATEGORIES AND FEATURES

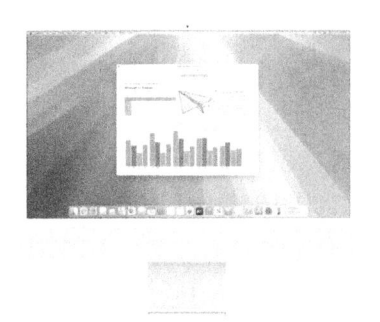

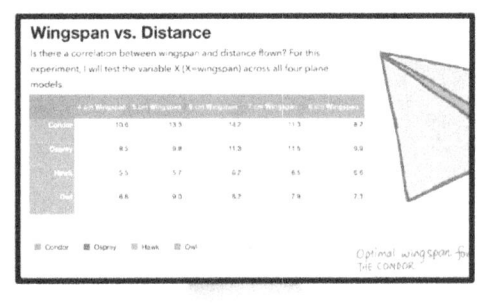

## 1. VISION

The Vision category offers features that improve the view display and the accessibility of the Mac interface for visually impaired users.

1) Zoom: Zoom is a feature allowing a user to zoom the screen display in or out for clarity. Zoom can be accessed via keyboard, by gesture, and via Accessibility Shortcut.

2) Display Options: Choose the preferred display - change to high contrast, adjust the brightness of the display, invert colors, or reduce motion as it may be easier on the eyes.

3) Pointer: The size enlargement, color change, and outline of the pointer will improve visibility; you can resize and change the color of the pointer to enhance it in different ways.

4) Color Filters: Color Filters alter screen color output for people experiencing color blindness and low vision. You have options like grayscale, red/green, green/red, blue/yellow, and many more.

5) VoiceOver: This screen reader reads aloud text into the ears of blind or low-vision users in order to work on the Mac. Through its settings, VoiceOver does allow for default adjustments to be made with regard to verbosity, sound, and Braille support.

## 2. HEARING

Hearing offers options to assist people who are hard of hearing, or can be used to enhance the user's experience when relying on audio cues.

1) Captions: Through text format, adjust the appearance of captions and subtitles. These can be used during videos and other forms of multimedia.
2) Hearing Devices: This feature can be used with compatible hearing aids or other devices to have the audio played directly, which can enhance clarity and customize the audio.
3) RTT: This feature will be very helpful for those who make/receive calls with texts. Real-Time Text allows you to type out texts during calls, and whatever you type will appear on both of your and the other person's screen in real time.
4) Live Captions: Real-time transcription of the spoken will go a long way in being helpful. The feature is useful for those who have a problem with hearing. The live caption allows modification for size, background, and font. This is available during video calls or during media playback.

5) AirPods Accessibility: On toggles the setting for accessibility features related to AirPods; these include changing the sound balance, optimized audio, and other settings that make AirPods more useable for those who have hearing impairments.

### 3. MOBILITY

Mobility settings provide options for those users who have limited physical or motor capabilities and thus need to control the Mac by methods other than standard input.

1) Voice Control: This is an option in which users control the Mac and applications by speaking commands. It is highly customizable, from nearly any aspect of the Mac.
2) Accessibility of Keyboard: In the event of a problem in normal typing, the facility of Sticky Keys and Slow Keys can be enabled. Sticky Keys: It allows key combination without necessarily pressing multiple keys simultaneously, whereas slow keys allows setting a delay before a key is accepted.
3) Assistive Touch Devices: For users who need alternative devices, a Mac can be connected to external assistive devices, such as joysticks, adaptive keyboards, and switches.
4) On-screen Keyboard: This feature allows users to type text and control the onscreen cursor without requiring a physical keyboard. The on-screen keyboard can be enabled to show up and hide when not needed.

5) Head Pointer and Facial Expressions: A user can configure their Mac to use head movements as a method to operate the pointer. Alternatively, face movements can be configured to insert a selected character, which can be helpful for those users with restricted hand movement.

## 4. SPEECH

The Speech section contains a set of features that include typing what you want to say and letting the Mac do the talking for you, thereby providing a voice for those who have defects in their speech.

1) Text to Speech: This feature allows the computer to read out loud the text typed. The type of voice, rate, and volume can be changed to suit particular settings.
2) Personal Voice: Users can create their own personal voice, based on their own, to be used for voice-increasing typed messages. This might add a comforting personal touch to digital communication. Speech Recognition: This feature allows users to dictate text rather than having to type it out. For people with limited mobility, or who simply do not like typing, this can be quite useful. The Mac will convert spoken words into text in real-time. 5. General
3) General: this section includes options to enable or disable features of accessibility quickly and it also provides options regarding keyboard shortcuts.
4) Accessibility Shortcuts: Configure your keyboard so you will be able to toggle on/off any

accessibility feature such as VoiceOver, Zoom, or any other tool without going to the menu.

5) Switch Control: By default, Switch Control is a form of accessibility that allows users to control their Mac using switches. The user will be working on on-screen items to enable them to select items, enter text, and control the Mac with an adaptive switch.

## TIPS FOR CUSTOMIZING ACCESSIBILITY SETTINGS

To effectively implement accessibility settings, here are some tips to consider:

1) Personalize to Needs: Fortunately, not all these features have to be necessary for every user. Customize settings in need of vision, hearing, or mobility, and speech capability to bring effectiveness in the use of Mac interface.

2) Take Advantage of Accessibility Shortcuts: To enable easier transitioning in settings, set up accessibility shortcuts. It reduces having to go deep inside menus, and in case one needs to turn things on/off, they can do it all in a snap.

3) Regularly Review the Preferences for Accessibility: Allow the settings to be changed and updated when needs evolve. This makes it easy to adapt the Mac interface for evolving accessibility requirements.

# CHAPTER 19: ADJUSTING YOUR VIEW WITH ZOOM

With the Zoom feature on your MacBook, you can magnify your entire screen or a part of it. Additionally, you will also be able to change the zoom level. Zoom lets you have closer views of texts, images, and controls without changing the display resolution on the Mac. It's pretty helpful for users who need to see what is on their screens up close.

## HOW TO SET UP ZOOM ON YOUR MACBOOK

To set up zoom using your computer:
1) Click on the Apple Menu. Then pick System Settings.
2) In the sidebar, click on Accessibility.
3) Tap Zoom from the available options. In this menu, you'll have a few different choices for how you want to use Zoom.

## KEY FEATURES AND OPTIONS OF ZOOM

1) Full Screen Zoom: This setting zooms the entire screen. You would want to use this option when you want the same zooming across all your screen aspects. You may select a higher or lower zoom level, depending on your preference, and go up or down to the next level as you see fit.

2) Picture-in-Picture: Split-screen zoom allows the selected portion of your screen to be zoomed out, while the remainder moves at normal resolution. This mode allows you to focus on a certain area while still being able to see what's going on with your whole screen.
3) Multi-Monitor Zoom: With this feature, one of the monitors-if using multiple-can be at a higher zoom while the other remains normal. This could definitely be ideal when one needs magnification on one display and referencing on the other.
4) Zoom Styles and Controls: You can create keyboard shortcuts for turning zoom on/off or increasing/decreasing zoom with a key click. For example, to increase the zoom, you press **Option-Command-Plus** (⌘⌥+); to decrease the zoom, press **Option-Command-Minus** (⌘⌥-).
5) More Zoom Options: Advanced settings enable the control of the movement of the area that is zoomed, whether the screen moves with the pointer, and which parts of the screen keep central.

These settings can be changed according to your preference to optimize the zoom in such a way that you see best but it doesn't interfere with reaching all parts.

## VOICEOVER – A SCREEN READER FOR MACOS

VoiceOver is the screen reader built-in on MacBook. It is for those who are blind or have low vision. It can read text from documents, the web, and system menus. Even images, app windows and controls are described. From this point onwards, VoiceOver lets you use your Mac with full accessibility by using just a keyboard or by using trackpad gestures. It also supports refreshable braille displays.

## HOW TO ACTIVATE VOICEOVER (OR DISABLE IT)

You can activate VoiceOver or disable it using any of the following methods:

1) Your Hot Key: **Command-F5** is the hotkey to enable and disable VoiceOver immediately. It is pretty useful for those users who like their screen to be read out immediately.
2) Via Siri: One can say, "Turn VoiceOver on" or "Turn VoiceOver off," and Siri does it accordingly.
3) Via System Settings:
   a) Open the Apple Menu, then System Settings.
   b) In the sidebar, click on Accessibility. To see it, one may have to scroll down.
   c) In the Accessibility section, select VoiceOver, then turn the switch to On or Off.

## KEY VOICEOVER FEATURES AND USAGE

1) Navigation Commands: Using VoiceOver will enable you to navigate your Mac with the use of different keyboard commands. For example, **Control-Option-Right Arrow** (^⌥→) will send you to the next item while **Control-Option-Left Arrow** (^⌥←) moves you to the previous

169

one. Learning some of these will even make navigation easier.
2) You can adjust the verbosity settings to show how much detail will be read out by VoiceOver. This includes punctuation, feedback on typing, and even more, all editable to adjust how much you want to hear.
3) VoiceOver speech provides several speech options, offering different voices along with speaking rates. Moreover, it offers facilities to add custom pronunciation or choose from a list of languages for better support of those users who use other languages primarily.
4) VoiceOver Fast Navigation and Gestures: With a Mac that includes a trackpad, the user has the ability to enable navigation by gestures. As an example, this covers moving left and right through items by swiping on a trackpad and performing an action on the selected item with a two-finger double tap.
5) Braille Display Compatibility: Along with the refreshable braille display, VoiceOver allows turning on the capability to let visually impaired users read Braille output from their Mac. In this respect, the integration ensures seamless support of Braille without third-party software.

VoiceOver does have some options for practice, where you'll be able to practice using various VoiceOver gestures and commands without having any effect on anything else within Mac. This will be helpful for learning to work with VoiceOver if you are using this feature for the very first time.

## ZOOM IN ON WORDS WITH HOVER TEXT

Hover Text is a feature for those users who want to see text in a larger format without having to zoom the entire screen. It will allow on-demand magnification of text beneath the pointer in a separate, zoomed-in window.

### CONFIGURING HOVER TEXT

To activate Hover Text:
1) Click Apple Men and then select System Settings.
2) Next select Accessibility inside the sidebar.
3) Select Zoom from the options list.
4) Scroll down and then turn on Hover Text.

### USING HOVER TEXT

Once the feature is turned on:

1) Press and hold the Command (⌘) key and hover the pointer over anything you want to see magnified.
2) A separate window will appear with the text in larger size and readability without distorting the main screen display.

### CUSTOMIZING HOVER TEXT

Hover Text can be customized to fit certain needs:
1) Window Size and Position: The Hover Text window is resizable and you can set where on screen it appears so you can position it for maximum convenience.
2) Text Size and Style: You will be able to specify font size, style, and background color for Hover Text in the Accessibility>Zoom settings.
3) Pointer Behaviour: You can set the pointer control option within your use of Hover Text for smoother and more natural movement across the magnified content.

Hover Text is good for its ease of use and efficiency on those who do not have to use text magnification enough, like reading documents or website text.

### ADJUSTING YOUR MAC DISPLAY COLORS WITH COLOR FILTERS

Color filters in macOS were for the vision impaired to make certain colors more distinct. These filters also help those who have visual discomfort or are sensitive to any color contrasts or brightness.

## SET UP COLOR FILTERS

To enable color filters:
1. Open Apple Menu and then pick System Settings.
2. In the sidebar, click on Accessibility.
3. Then click on Display and select Color Filters.
4. Tap the switch located beside Color Filters to switch it on.

## SELECT AND CONFIGURE COLOR FILTERS

Once color filters are turned on, you have a number of options available to you:

1) Gray scale: This turns screen colors into different shades of gray, which for those sensitive to color may be easier to distinguish.
2) Red/Green Filter (Protanopia/Deuteranopia): This filter allows users who have Protanopia or Deuteranopia (red-green color-blind) to distinguish between colors that might seem too similar.
3) Blue/Yellow Filter (Tritanopia): This filter assists users with Blue-Yellow Color-Blindness.
4) Custom Tint: You are allowed to choose a certain tint to apply for the screen. This serves a couple of purposes - primarily, glare can be reduced and light sensitivity for users is enhanced.

## COLOR FILTERS QUICK TOGGLE

To immediately enable or disable color filters:
1) Press Option-Command-F5 to open accessibility shortcuts that include color filters.

2) On Macs / Magic Keyboards with Touch ID, you can access the Accessibility Shortcuts menu by quickly pressing Touch ID three times and then immediately be able to turn color filters on/off.

Color filters are more flexible in terms of color customisation; hence, it allows the creation of a display environment that best suits the visual requirements of the user.

# CHAPTER 20: HEARING ACCESSIBILITY FEATURES – LIVE CAPTIONS

Live Captions automatically generate the text of spoken words in real-time for conversations, audio, and video playback, making audio information accessible to users who are deaf or have low auditory sensations.

## HOW TO ENABLE LIVE CAPTIONS

To setup Live Captions:
1) Open System Settings from the Apple Menu
2) Select Accessibility from the sidebar
3) Scroll down to Hearing, then click Live Captions
4) Enable Live Captions

## SOME KEY FEATURES OF LIVE CAPTIONS

1) Real-Time Transcriptions: Live Captions generate captions for audio played on the Mac, including video calls, movies, and online videos. These are then displayed in a window for easy readability of, and to follow, the caption.
2) Speaker Identification: Where possible, Live Captions identify who is speaking to make it easier to track multiple conversations.

3) Appearance Options: You can adjust font size and color and background color in the caption window to make things more visible and easier to read.

## LIMITATIONS OF LIVE CAPTIONS

It is very critical to note that:
1) Availability: Live Captions are not supported in all languages, regions, or countries. If Live Captions are available based on your location, it may need some checking within your settings.
2) Accuracy: Much as they prove helpful, sometimes Live Captions may not be fully or always accurate. This may happen especially in very noisy environments, complex audio, or during specialized vocabulary. Apple, Inc. encourages you not to rely exclusively upon Live Captions in high-risk or emergency situations.

Live Captions provide an essential accessibility feature, ensuring greater inclusivity and access to multimedia and conversational content by people with hearing impairments.

## PAIR HEARING DEVICES WITH YOUR MAC

If you use MFi hearing aids or sound processors (MFi means Made for iPhone), you can pair them with the MacBook to stream audio and change device settings for a more personalized hearing experience.

## HOW TO PAIR HEARING DEVICES

1) Apple Menu: Click the Apple menu at your screen's top left.
2) Open System Settings: Select System Settings from this menu.
3) Go to Accessibility: Access this from the sidebar by clicking on Accessibility.
4) Select Hearing Devices: Further down, select Hearing Devices. This is the section where any MFi hearing aids or sound processors can be paired.

### ONCE PAIRED:

1) Audio Streaming: Your Mac can stream system sounds, media playback, calls, and other audio directly to your hearing device.
2) Device Adjustments: The Accessibility settings let you control volume adjustments, sound level, and turning specific hearing modes on if your device supports these features.

Pairing MFi hearing aids, users are able to enjoy a more integrated audio experience on the Mac with even clearer sound and more precise control over what they hear.

## CUSTOMIZING AIRPODS FOR HEARING ASSISTANCE

AirPods have integrated accessibility features designed to amplify the sound and adjust specific audio frequencies with the aim of just making music, movies, calls, and podcasts sharper and clear. This personalization will surely relieve listening to audio more with full details and clarity from people with hearing difficulties.

**How to Customize the AirPods**
1) Open System Settings: Apple Menu and then pick System Settings.
2) Click on Accessibility in the sidebar.
3) Open the Settings for AirPods: In the Hearing section, you find either AirPods Settings or Audio- the naming depends on which version of macOS your MacBook is using.

**Some of the Most Important Accessibility Settings in AirPods**
The section for AirPods allows for:
1) Audio Amplification: Amplify soft sounds to make them easier to perceive while keeping louder sounds unchanged.
2) Adjust the Frequency: Choose the exact frequency range that you want to amplify; this serves to fine-tune the audio output with your individual needs.
3) Transparency Mode Settings (AirPods Pro): Adjust how outside sounds mix in with your audio for better awareness of your surroundings while listening to active AirPods.

These are very helpful settings for those who may require more clear audio in certain situations or environments. Adjusting the settings on one's AirPods allows for personalized listening based on one's needs or requirements in hearing.

# BACKGROUNDS SOUNDS TO MASK AMBIENT NOISE

The MacBook Background Sounds feature provides a selection of relaxing audio options to help drown out distractions or mask unwanted environmental noise for when these noises need to be heard. It's a tool designed to help you pay attention or relax anywhere.

**How to Turn On Background Sounds**
1) Open System Settings: From the Apple menu, select System Settings.
2) Click on Accessibility: Select from Accessibility in the sidebar.
3) Click Audio: Scroll down and click on Audio settings.
4) Enable Background Sounds: Tap to enable Background Sounds.

**Playing and Customizing Background Sounds**
When Background Sounds is on, you can do the following:
1) Select a sound type: Tap to select Rain, Ocean, or Stream to hear a soothing sound that masks background noise.
2) Adjust Volume: Utilize the volume slider to determine how loud the background sound should be, compared to other audio, such as music or podcasts, to set the background sound for your needs.
3) Continuous Playback: Optionally configure the background sounds to play continuously, even

when other media is playing, to maintain a constant noise mask and help you focus better.

Background Sounds are excellent, for they minimize distraction and reduce anxiety by creating an environment that is comfortable to listen to, hence more productivity and better relaxation.

# CHAPTER 21: VOICE CONTROL ON MACBOOK

This feature allows users to perform numerous activities with their voice command. It has been designed to listen to natural speech, guiding users of Mac through the interface, launching applications, and typing text, among other things. Importantly, all audio processing for Voice Control happens locally on the device for preserving user privacy.

**How to Turn On Voice Control**
1) Tap Open System Settings: Click the Apple menu at your screen's top left, pull it down, and click System Settings.
2) Tap Accessibility Options: In the sidebar, click on Accessibility.
3) Activate Voice Control: Scroll down to click Voice Control to turn this setting on.

**Main Components of Voice Control**
After turning on Voice Control, you can do the following
1) Navigate the UI: Using Voice Control, you could activate applications, navigate and select menus, and even modify settings by saying something like "Open Safari" or "Click System Preferences. "
2) Dictate Text: You could say something like sentences in text fields, and Voice Control would type it for you; even punctuation will be handled if you include that in your command.

3) Create Custom Commands: The Voice Control feature does support custom voice commands in order to do common operations far more easily than ever before.
4) Apply Numbered Overlays: If a layout is too complex, then Voice Control can turn on and pop up an on-screen grid or numbers overlay for easier selection of target elements by their numbers rather than having to use the name.

Using Voice Control, you would be able to work effectively on your MacBook hands-free for accessibility and convenience that users with mobility impairments may have.

## PERSONALIZE YOUR POINTER FOR EASIER VIEWING

For easy locating and tracking of the mouse pointer, MacBook gives you an opportunity to change the outline and the fill color of the pointer. This makes it easier for your eyes to track the pointer whenever it changes shape or moves across high-resolution screens.

**How to Personalize the Pointer**
1) Open System Settings: Apple Menu and then pick System Settings
2) Go to Accessibility: Inside the sidebar, tap on Accessibility.
3) Select Display Options: Scroll down to the option Display, where you get the option to change the settings related to the Pointer.

**Pointer Options**
After pointing to the settings in the Pointer, you are allowed to:
1) Change Outline Color: You can change the color of the outline of the pointer so that it stands against all kinds of backgrounds. It will be helpful for those users also who require additional contrast.
2) Change Fill Color: Change the fill color of the pointer to enhance the visibility and customize it according to your preference.
3) Change Pointer Size: MacBook also allows changing the size of the pointer by making it bigger in order to see it more comfortably.

These modifications also help locate and track easier, especially to those who may have limited difficulties in locating or following pointers of size smaller than conventional or usual forms.

## 3. BETTER KEYBOARD ACCESS FOR FULL CONTROL

MacBook provides extended keyboard accessibility with which users can control almost everything onscreen without using a mouse or trackpad. One can forward and backward, navigate, open apps, and even perform interface elements like buttons and sliders with the use of shortcuts.

**How to Turn On Keyboard Access Features**
1) Go to System Settings: Open the Apple Menu and then pick System Settings.
2) Go to Accessibility: Inside the sidebar, tap on Accessibility.

3) Enable Full Keyboard Access: Scroll down to locate Keyboard options, where you will enable Full Keyboard Access and customize the settings.

**Key Capabilities of Extended Keyboard Access**

If you have keyboard access on, then:
1) Ranked Interface: You will use the Tab and Arrow keys to move among elements on the screen. These may include buttons, sliders, or even menus.
2) To Activate Elements: Press Enter in order to select an activated item. Examples are application icons, settings, and even dialog boxes.
3) Customize Shortcuts: Design your own shortcuts to perform specific actions, therefore, further extending control over the interface.
4) Modifier Keys: In addition to the main modifier keys, Control and Option extend the types of actions you can carry out via keyboard and, consequently, make it easier to interact with complex menus or special functions without toggling to a mouse.

Full Keyboard Access is designed for those who like or need hands-free control. You can have the most complete MacBook control using your keyboard only.

## HAVE YOUR MAC SPEAK FOR YOU WITH LIVE SPEECH

With MacBook's Live Speech, users can type a message and hear it spoken. This can be very helpful for those struggling with-and in-turn prefer the assistance of-speech through text-to-speech functions. It is available on FaceTime calls, regular phone calls, and even in-person conversations, making this feature have many different uses.

**How to Turn On and Use Live Speech**

1) Open System Settings: From the top left corner, select the Apple menu and click on System Settings.
2) Head to the Accessibility: In the sidebar, select Accessibility.
3) Click on Speech: Among the options, click on Speech to go to the settings for Live Speech.
4) Enable Live Speech: Turn on Live Speech now.

**Key Features of Live Speech**

Once turned on, there are the following options available:

1) Type-To-Speak: In the call/conversation sessions, there is a small input field in which you type your text message. It automatically speaks out as typed when submitted by the Mac.
2) Save Common Phrases: You can save commonly used phrases like greetings and responses for later quick access. This is very helpful during calls or conversations, as clients can then easily "chime in" without them having to retype similar text.
3) Voice Preference: This allows you to select a preferred voice in which the speech output should be produced. Apple provides multiple voices with

185

various accents and tones; hence, you can select any one of them with which you feel comfortable.
4) Record a Personal Voice: Apple makes it possible to record a Personal Voice, allowing users to save a version of their voice for Live Speech; this really adds a much more natural and personal touch to conversations.

Live Speech is useful in maintaining the continuity of conversations, hence very handy for those users who want to take an active part in verbal interactions without speaking directly.

## USE VOICE SHORTCUTS FOR INSTANT ACTIONS

Using Voice Shortcuts, one can get certain activities done by just saying their keyword/phrase/sound. You can configure your Mac to connect your voice to specific actions like opening apps, launching any particular process, or turning any setting on and off. Thus, this is a fairly useful option if you want to be able to control your computer hands-free.

### HOW TO CONFIGURE VOICE SHORTCUT

1) Open System Settings: From the Apple menu, select System Settings.
2) Accessibility: Click on Accessibility in the sidebar.
3) Find Speech followed by Vocal Shortcuts. Click to open the settings for Vocal Shortcuts.
4) New Shortcuts: To create new vocal shortcuts, type out the word, phrase, or sound you want to

begin an action of your choice. The system will prompt you for what action and vocal command
  a) Custom Voice Commands: Set up custom voice commands to open any application, launch a website, or initiate any action in your Mac.
  b) Sound Sensitivity Settings: Different speech tones and volumes may require setting the sensitivity to fine-tune it for appropriate recognition of your Mac to hear your commands.
  c) Pre-set and Custom Commands: Along with pre-set commands, one can set up customized commands so that more flexibility can be ensured in terms of control over how far one wants to go using his/her voice with a Mac.

This is a much welcome feature for those who may find navigation through conventional means tricky or would simply prefer voice-activated control.

## LET SIRI LISTEN OUT FOR ATYPICAL SPEECH

If one has a condition that affects their speech, like an acquired or progressive impairment of speech, then one can modify Siri to understand atypical speech patterns. This will help Siri recognize and process voice commands more precisely. This feature makes the product more inclusive.

**How to Configure Siri to Listen for Atypical Speech**
1) Open System Settings: Apple Menu and then pick System Settings.

2) Go to Accessibility: In the sidebar, select Accessibility.
3) Tap on Speech: From the list of options under Accessibility, find and tap on Speech.
4) Enable Atypical Speech Recognition: Scroll through and look for the option that will make Siri listen for atypical speech patterns; toggle this on so that it can enhance Siri's capabilities in capturing nonstandard speech.

**Features of Siri's Atypical Speech Recognition**

1) Enhanced Speech Recognition: Siri gets used to more speech patterns and sounds, hence access by people who have quite different or changing speech features.
2) Reduced Incidents of Errors in the Interactions: Because of tuning to atypical speech, there's less likelihood for misunderstandings in what should be done, hence an overall smoother user experience.
3) Continuous Learning: Siri keeps getting even better, as with every interaction accuracy and responsiveness improve to specific speech nuances.

This feature allows Siri to support users with different speech patterns, thereby making the MacBook more usable and accessible.

# CHAPTER 22: MANAGING WINDOWS

Window management on MacBooks is key to keeping your workspace clean and organized, especially when working with a lot of applications that may require several windows open simultaneously. Here are the major tools and techniques in macOS that make window organization efficient:

## PREVIEW

### FULL-SCREEN MODE FOR FOCUS ON A SINGLE APP

Full-Screen Mode lets an app take over the entire screen to help minimize distractions from other open apps and lets you focus on a closer view of your work. This is beneficial for projects that are very immersive, such as writing, video editing, or graphics design.

**How to Enter Full Screen Mode**

1) Open an application you wish to use in full screen mode.
2) Tap the green button with two arrows in any application window's top left corner or hover your mouse over it and select Enter Full Screen.
3) The application now fills your entire display while other applications disappear from view until you return to this view.

## Exiting Full Screen Mode

Take your mouse over to the screen's top to see the menu bar, and then tap the green button again, or hit Control + Command + F.

### SPLIT VIEW FOR MULTI-TASKING WITH TWO APPS

Split View ideal when you want to work with two apps right beside each other, like referencing a document you have typed while taking notes, or having the web open beside a video.

## How to Use Split View

1) Open the first application, hover your cursor over the green full-screen button and select either Tile Window to Left of Screen or Tile Window to Right of Screen.
2) Open the second app and it will automatically fill in the screen's adjacent side.

## Adjusting the Split

To resize the split, drag the vertical line between the two windows to give one app more space.

## Exiting Split View

Click to close one of the apps or click the green button to exit full-screen mode; either will collapse Split View and return both apps to their separate windows.

## STAGE MANAGER FOR AUTOMATIC WINDOW MANAGEMENT

Stage Manager is an option for auto-arrangement of open apps on your desktop. Because windows are grouped together by default, Stage Manager reduces desktop clutter and makes it easy to switch between groups.

**How to Activate Stage Manager**

1) Click the Apple Menu and then pick System Settings.
2) Click the Desktop & Dock inside the sidebar.
3) Put on Stage Manager to have your desktop automatically organized.

**Using Stage Manager**

1) Active App: The active app window is brought in sharp focus in the center of your screen.
2) Grouped Windows: Other open apps or windows are minimized onto the left sidebar. You can click these to bring them up to the foreground.
3) App Switching: Click any app in the sidebar to bring that into the main view, replacing the current app. This setup is perfect for users who want a number of different apps open but have a streamlined way of toggling between them without cluttering the main workspace.

## MISSION CONTROL FOR QUICK ACCESS TO ALL OPEN WINDOWS

Mission Control unleashes a layer that shows you all of your open windows, desktops, and full-screen apps in an organized overview, so you can look at everything running on your Mac in one instant. And it is quite helpful when you need to find that one window or app buried underneath all others.

### How to Open Mission Control
1) Scroll up with three fingers on your trackpad, or press your keyboard's F3 key (Mission Control key).
2) All open windows will appear in a tiled view as well as other open spaces and full-screen applications will appear at the top.

### Using Mission Control
1) To bring any window to the foreground, click on it
2) To add a new desktop space hover over the top then click the "+" button on top.

## MULTIPLE DESKTOP SPACES (VIRTUAL DESKTOPS)

Multiple desktop spaces, called "Spaces," let you set up virtual desktops for various activities or projects you're working on and then easily switch between them. For example, you can have one space with work applications open and another space with personal things.

### Creation and Spaces Switching
1) Open Mission Control described above.
2) In the top bar, click the "+" sign to add a new desktop.

3) To switch between spaces, three-finger swipe on your trackpad left or right or use Control + Left/Right Arrow

**Moving Windows Between Spaces**

Drag one window from one space to the other at the top of the screen in Mission Control

Multiple spaces are very good to split your work depending on different projects or workflows; you are going to be able to keep all of your stuff organized and jump between contexts really fast.

## USING FULL-SCREEN VIEW ON MACBOOK

Full-Screen View opens one application window to fill the entire display screen, so that you don't see any other applications, the Dock, or the menu bar; you work only with the active application.

**How to Enter Full-Screen View**

1) Open the application you want to work in.
2) Click the green button at the top-left side of your application window; it contains two arrows.
3) Hover the green button, and from its drop-down menu select Full Screen - or simply click on the button.

When you go into full screen, your whole screen will be accommodated by the app, covering everything else running and background distractions.

**How To access the Menu Bar and the Dock which are Hidden:**

1) Menu Bar: Just take your pointer to the top of the screen; the menu bar will reappear.
2) Dock: In case your dock is automatically hidden, you only have to move your pointer towards the bottom of the screen. You can also set your preference to not auto-hide the dock in System Settings > Desktop & Dock.

**How to Quit Full-Screen Mode**
1) Viewing green button, move cursor at the top of the screen.
2) Green button - Click to Quit full-screen mode OR press Control + Command + F.

**Always Show Menu Bar:**
You can change settings to let the menu bar appear during full-screen mode. To do so, follow these steps:
1) Open the Apple menu, click System Settings, and proceed to click Desktop & Dock.
2) Turn off the option to *make the menu bar disappear and reappear when you go full screen* so that it remains always visible at times when in full screen for easy access to tools.

## USING SPLIT VIEW ON MACBOOK

With Split View, you can work with two app windows side by side, each taking half of the screen. This really comes in handy with those working projects when you need to refer to information from one app while working in another-appropriate for such tasks like researching and writing, or comparing documents side by side.

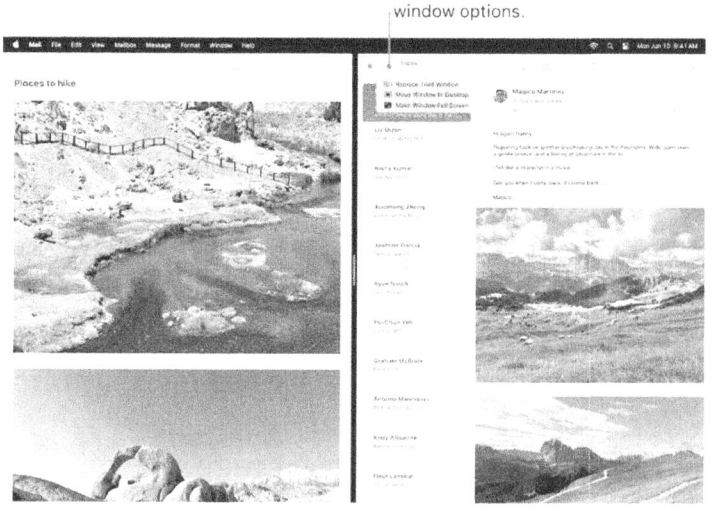

## **How to Turn On Split View:**
1) Open the first app window that you want to set up with Split View.
2) Once the app window is active, to move the tile window to the left or right side of the screen, place your mouse pointer over the green button located in the window's upper left corner and select the desired option.
3) The window will jump to the half of the screen you chose while other currently open apps are listed as choices for the opposite half.
4) Click the other app window you want on the opposite side of the screen, and it will automatically fill the remaining half of the screen.

Once in Split View, each app occupies one-half of the screen, and you can work in both windows without having to manually resize them.

### Resizing Windows in Split View:
1) A vertical bar divides two app windows when in Split View.
2) Drag this bar to set the relative amount of screen space given to each app. This is helpful if one application requires more focus than another.

### App Switching and Quitting Split View:
1) In order to replace one of the applications in Split View, hover over the green button; other options will come up, and you can replace the application in one half of your screen.
2) To exit Split View, click the green button of one of your app windows, or press Esc. Both windows will return to their normal size.

### Full-Screen Mode of Each App in Split View:
If you want to focus on one app from the Split View, you can go full screen by positioning your cursor at the screen's top and choosing Full Screen from among the options given through the green button.

## FULL SCREEN VS SPLIT VIEW:
1) Full: Good for focusing on a single app without distractions
2) Split: Good for multitasking in general: using two apps side-by-side without clutter of more windows

## TILING WINDOWS ON MACBOOK

The tiling of windows allows you to organize your applications on the screen without any overlap and hence easily see several documents or programs all at once. Here is how it is done.

**How to Tile Windows**
1. Open Application: Open all applications or windows that you want to tile on your MacBook.
2. Locate Green Button: The green button is located near the top-left of the window. It consists of two arrows pulling in opposite directions. This is the same button used to make things full-screen and now is used for tiling.
3. Enable Tiling: Bring your pointer over the green button. The drop-down opens showing options for tiling. Select Your Arrangement:
4. Tile Window to Left of Screen: This will move the selected window to the left half of your screen.
5. Tile Window to Right of Screen: This will position the window on the right half of the screen.
6. Once one of these is selected, the window will snap to that side of the screen.
7. Click on the Second Window: After having one window tiled, the rest of the open windows will show upon the other half of the screen. Click the window that you want to occupy another half of the screen, and it will automatically fill that half of the screen.

**Resizing Tiled Windows :**
1. Window Resizing: By dragging the dividing vertical line separating the two windows in tiles,

you will be able to resize windows. This will give more screen space to the application that you are working on.
2. Tiled Apps Switching: In order to switch the running application in one of the positions, just hovering over the green button of another window you want to switch to and clicking on the tiling option again will do the deal.

## MORE WINDOW MANAGEMENT TIPS

### Clean Up Your Desktop in a Jiffy:

On occasions, opening a lot of windows could crowd your desktop. You might want to work or open some files on the desktop, so follow these steps when it comes to clearing the view, easily and at once:

1) Click on the Desktop Wallpaper: Press anywhere on your desktop wallpaper. This automatically minimizes all the windows that have been opened up and you can clearly see the desktop itself.
2) Reopen Your Windows: To restore your windows, click on the desktop wallpaper again. All the previously minimized windows will appear to the last position, and you can easily carry on with your work.

### Managing your Windows with Mission Control:

1) Mission Control: It is another useful feature that allows the user to view all the currently open

windows along with open spaces all at once. To use Mission Control:
a) Swipe three or four fingers up on your trackpad or press the F3 key-or at least the key with three rectangles on the top row of your keyboard.
b) It provides an overview of all applications opened and their windows to select the one you want more easily.

## USING SPLIT VIEW FOR EXTEND

While tiling windows is great for side-by-side viewing, try Split View to take multitasking to the next level below:

**How to Turn on Split View:**
1) Tile one window using the previous steps
2) Select a second window to open to fill the remaining part of the screen.
3) Split View allows both applications to be viewed and engaged without overlap.

## WIDGETS ON YOUR MACBOOK DESKTOP

Widgets are small applications that let you access information or perform day-to-day tasks quickly and all from your desktop without consuming too much space. Here's how to add, remove, and rearrange widgets:

1) Managing Your Widgets: To do this, click the time and date in your menu bar at the top of the screen. This opens the Notification Center.
2) Alternatively, right-click anywhere on the desktop background and click Edit Widgets from the context menu.
3) Adding Widgets: With the widget gallery now opened, you will see different varieties of widgets available. These may include events in your calendar, weather information, reminders, and others. Browse through the available widgets. Each widget may offer different sizes or options to display specific information.
4) To add a widget, simply drag it from the gallery to your desktop or Notification Center. You can place it anywhere you like, customizing your layout based on your preference.
5) Removing Widgets:
    a) To remove a widget from your desktop, hover over the widget until you see the remove button appear-usually a small "x" or a minus sign.

    b) Use the remove button to delete the widget from your desktop.
6) Rearranging Widgets:
    a) You can reorder widgets by clicking and dragging them where you'd like, so you can work in a layout that best suits your workflow.
    b) To resize some widgets, you may find options either in the widget gallery or on the widget itself for showing small, medium, or large sizes.
7) Adding iPhone Widgets: If you are signed into one Apple ID on both your iOS device and Mac, then you can directly add the widgets from your iPhone to the Mac desktop. This serves well for continuity whereby you will have access to your favorite iPhone widgets without having to install corresponding apps into your Mac.
8) Customization Of Widgets: Once you have set up your widgets, it's often possible to personalize their look - theme or data to display, for example - by clicking on the widget itself and delving into its settings. This feature lets you personalise the information within the widget for better optimisation of it.

## STAGE MANAGER

Essentially, Stage Manager is a feature designed to keep tidiness on your desktop. It does this by putting your opened apps and windows in order automatically, keeping clutter minimal but still keeping your active tasks at hand. In this tutorial, we're going to give you a run for your money, showing how to work effectively with Stage Manager. Here's how to utilize Stage Manager:

### Enable Stage Manager
1) Open the Control Center: Begin by tapping the Control Center icon in your menu bar; it has the icon of two switches. This opens a group of options for the management of system settings.
2) Then enable Stage Manager: Find Stage Manager in Control Center and tap to activate it. Inactivated, the desktop will be organized to focus on the application currently being used front and center, while other open windows organize neatly on the side.

### The uses of Stage Manager:
1) Focusing on one app: When an application opens, it opens right in the center of your screen. All the rest of the applications slide to the side so that you can work on whatever you want without any kind of clutter.
2) Other Windows: To switch to another application or window, simply click on it from the side setup. It brings the window to the front, and hence, toggling between tasks becomes quite easy.
3) Customize Your Stage Manager Setup: Drag windows on and off the side to set how many apps show up in Stage Manager. And if you have several desktop or space, Stage Manager remembers your setup across your spaces.

**Benefits Of Using Stage Manager:**
1) Reduced Clutter: By organizing all of your windows and apps, Stage Manager helps you create a cleaner workspace. And the fewer windows to distract you, the better.
2) Improved Accessibility: All your regularly used applications are just a click away and hence no more difficulty in multitasking, which means better workflow surely.

## MANAGING WINDOWS AND DESKTOPS ON YOUR MACBOOK WITH MISSION CONTROL

Mission Control is a powerful feature on macOS that extends your workspace by showing you all opened windows and handling multiple desktop spaces, also called "spaces." Mission Control gives you a quick glance at your applications; it keeps your workspace cleaned up and makes switching between applications easier. Mission Control lets you have multiple virtual desktops. Here's how you can use Mission Control and how to create and manage multiple desktops using it.

**What Is Mission Control?**
Mission Control shows all opened windows, split views, and spaces in one view. You can locate right away and switch to any application. Once you enter Mission Control all active windows show up in one layer so that you don't get lost in task switching.

### Entering And Exiting Mission Control

You can turn on Mission Control the following ways:
1) Keyboard Shortcut:
   a) Press the F3 key - those three little rectangles ▢▢ at the top of your keyboard. It's usually labeled with the Mission Control icon.
   b) You can also hit Control + Up Arrow for Mission Control.
2) Trackpad Gesture: If you're using a MacBook or a trackpad, swipe up using 3 or 4 fingers upwards depending on how you have set up your trackpad, to enter Mission Control.
3) Dock Icon: You can also add the Mission Control icon ▦ to your Dock for quick access. To do so, first open your Mission Control, right-click the desired icon ▦ in the Dock, and select Options > Keep in Dock.

### Using Mission Control

Immediately after entering Mission Control, you will see all the opened windows are arranged in one layer. Here's how to navigate and make full use of this feature:
1) Choose a Window: Clicking on any window brings it to the front and active. There is no need to minimize any of the windows or attempt to reach a window that may lie buried beneath.
2) Viewing Multiple Spaces: If you created any additional desktops, or spaces, those will be

displayed, as well as the apps you're using in Split View. You can click the top of the Mission Control screen on any of these thumbnails to jump immediately to that particular space.

3) Close or Rearrange Windows: You can also drag windows around in Mission Control to move them, or close them by hovering your cursor over the window and clicking the close button - the red dot.

## CREATE AND MANAGE DESKTOP SPACES

If one desktop isn't sufficient to keep your work organized, you can create multiple desktop spaces using Mission Control. As a matter of fact, this is a great way of organizing your work across several projects or tasks.

**Creating A New Desktop Space:**

1) Enter Mission Control: Clicking the + icon, or by pressing Mission Control: If you followed the steps above to enter Mission Control.

2) Add a Space: Place your mouse cursor to the screen's top right-hand corner, and click the + button-this means "Add Desktop." This opens a new desktop space.

3) Name Your Spaces: By default, new spaces will be named "Desktop 1," "Desktop 2," etc. You can right-click on the space and Rename for more natural naming that is easier to identify in later steps.

**Switching Between Spaces:**
Once you have set up several, it becomes second nature to change between them using the following ways:
1) Keyboard Shortcuts: Control + Right Arrow and Control + Left Arrow keyboard shortcuts will enable you to flip between desktop spaces.
2) Mission Control: Just jump into Mission Control and then click at the top of the screen on which space you'd like to flip to.
3) Trackpad Gesture: With a trackpad, swipe left or right using three fingers to switch across spaces (you can also swipe with four fingers).

**Moving Windows Between Spaces:**
You can drag windows between other spaces to better arrange them:
1) Activate Mission Control.
2) Drag a Window: Click and drag the window of your choice to the top of your screen to your desired space by holding onto that window. Let go of the mouse to drop that window in that space.

### DELETING SPACES

To delete any space you no longer need:
1) Go into Mission Control.
2) Hover Over the Space: Hover your cursor over the space which you want to delete. You will see an "X" appear in the corner of the thumbnail of the space.

3) Space Removal: Click the "X" to remove the space. Any open window in that space will move to another available space.

## MISSION CONTROL-USAGE BENEFITS

1) Clarity and Focus: this feature allows you to operate multiple spaces and dedicate distinct screen portions to whatever is at hand, therefore reducing or completely avoiding diversion of attention.
2) Quick Access: through Mission Control, you'll get access to any opened application in the shortest time; this may save you from wasting much time in search of windows.
3) Organization of the workflow: Creating separate desktop spaces makes your workflow organized if you are working on multiple projects or more than one task simultaneously.

## THE RED, YELLOW, AND GREEN BUTTONS

**Red Button - Close Window**
The red button mainly closes the current window of any application that you are using. However, the result for clicking this button can be different depending on the application you are working with:
1) Close Current Window: This only closes the current window in most of the applications. The application, however remains open and you can

use it as you want. Using Safari, if you click on the button in red, the window of Safari will be closed while the app stays opened and you may open another window or tab later.

2) Quit Application: For some applications - especially utilities or lighter programs - this red button will quit the whole application, therefore closing all windows. Examples of applications that would do this are TextEdit or Preview. For these programs, clicking the red button exits you completely out of the application. Tip: You can actually just quit an application without having to close the window by using Command + Q.

## **Yellow Button (Minimize Window)**

The yellow button minimizes the current window, temporarily hiding its view. When you minimize a window, it slides down into the Dock on the right side.

1) Temporary Closure: The minimized window does not get closed; instead, it shrinks down to an icon in the Dock. This helps the user clear his/her desktop without losing work or the state of the application.

2) Restoring a Minimized Window: To open a minimized window, click the icon of the window in the Dock. The window opens to its previous size and to its previously set location.

**Note**: Press the Command + M keys to instantly minimize the currently active window.

## Green Button for Full Screen and Split View

The green button has a number of functions, foremost among which is that it makes the window switch to full screen or allows for the activation of Split View thus enabling you to work with two windows side by side.

1) Full Screen: The green button opens the current window full screen, without seeing either the menu bar or the Dock. To exit full-screen mode, move your cursor to the top of your display and click the green button in the top-left corner again, or press Esc.
2) Split View: To work with two different applications side by side, click and hold the green button. This will bring up options to place the window either on the screen's left or right side. Once you have selected a side, you can choose another open window to fill in.

**Note**: For quick toggling between full-screen and windowed mode, double click the title bar of the window or use the key combination Control + Command + F.

## Making The Most Of The Traffic Light Buttons

To fully utilize the traffic light buttons, then here are some things you could try doing:

1) Know the Application Behavior: At times, certain applications shut completely down when the red button is clicked. Many of them do not. Know what an app does, and that aids in workflow management.

2) Maintain Your Desktop: Use the yellow button to minimize windows instead of closing them. Keep your desktop organized; this helps you get to any application easily, without having to reopen it.
3) Split View for Multitasking: Use the green button to multitask in Split View. Perhaps you want to compose an email as you have a document open to your side; this is very helpful if you are doing research, writing, or need to have more than one application on at the same time.
4) Keyboard Shortcuts: Also, memorize the keyboard shortcuts for these actions: Command + W for close, Command + M for minimize, and Control + Command + F for full-screen to have smooth work and save your time.

# CHAPTER 23: MAC OPERATING SYSTEM REINSTALLATION

The most useful thing you can do to troubleshoot problems of the system, upgrade to a clean install, or reset the system without removing your personal files, is the reinstallation of macOS in an MacBook via macOS Recovery. A step-by-step guide for how to carry out this procedure is given below:

## STEP 1: REINSTALLATION - PREPARATION

While it won't erase any of your personal data or applications, it is quite sensible to back up your Mac just in case anything might go wrong along the way when trying to reinstall macOS. Use Time Machine or whatever else you feel comfortable with to make certain that your files are safe, should anything go wrong during the process.

1) Back up with Time Machine: Connect an external drive and, through the **System menu**, go to **General > Time Machine**
2) Click **Add Backup Disk** and follow the on-screen steps.

## STEP 2: TURNOFF YOUR MAC

To perform the reinstall, you will need to have your Mac fully powered off.

1) Shut down with Apple Menu:
   a) Click the Apple menu  in the top left corner of the screen.
   b) Click on Shut Down.
2) Force Shut down - Using Keyboard Command
   a) If your Mac becomes unresponsive and won't turn off, press the power button and hold for 10 seconds to forcibly turn it off.
   b) For Macs that include Touch ID, press and hold the Touch ID button until the computer turns off.

## STEP 3: START FROM MACOS RECOVERY

This involves starting up your MacBook from the macOS Recovery. It helps you access those tools necessary for the reinstallation of macOS, restoration from a backup, or troubleshooting of your computer. It comes with a few options; hence, one can reinstall the most recent version of macOS which is compatible, use the Disk Utility for the purpose of verification or repair of disks, restore from a Time Machine backup, or access online support tools. When serious software issues occur, or when a fresh install of macOS is at hand, macOS Recovery serves as a great way to repair or reinstall the OS.

## STEPS TO STARTUP FROM MACOS RECOVERY ON APPLE SILICON

The Mac running with Apple silicon, such as the newest M1, M2, M3, and M4 chip models, differ slightly due to the beginning process; hence, steps differ a little. How is it done? Well, let us go into detail:

1) Press and Hold the Power Button
   a) Shut down your Mac if it is currently turned on by choosing **Shut Down** from the Apple menu or by pressing the power button and then holding it until the Mac turns off.
   b) Press the power button - also, on some models, the Touch ID button - and hold them down until the computer powers up.
2) Keep Holding the Power Button: While you keep pressing the power button, the Mac will turn on but immediately boot into a mode where it directly loads Startup Options, instead of loading directly into the macOS operating system. This may take some seconds, so hold the button down until you come to the Startup Options window.
3) Startup Options Window:
   a) Once the Startup Options window appears, select among other options, the internal storage drive with a button labeled Options.
   b) When it boots, release the power button. You will see options appear to enter Recovery mode. Sometimes you may also see a list of all the available disks if multiple drives are attached.

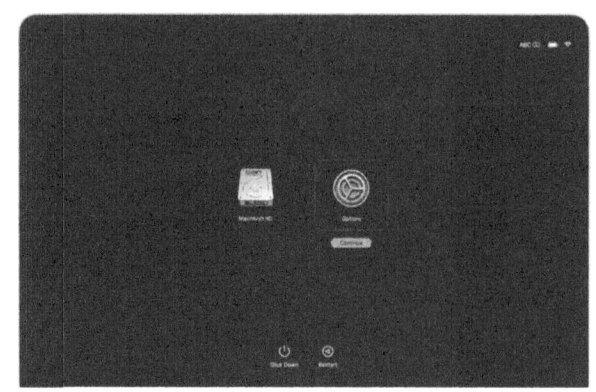

4) **Click Options and Continue**: Click Options; this will open macOS Recovery. Then, click the Continue button below it. This will start loading the utility of macOS Recovery that contains various tools for reinstalling or repairing macOS.
5) **Authenticate with Admin Account if required**:
   a) When macOS requires authentication, and will prompt you to do so using an administrator account, select a user account that has administrative privileges (main account, usually) then click **Next**.
   b) Enter the password for the selected administrator account: You will be putting in the password used to log in with the Mac during startup. This is a security step toward ensuring that only authorized users access the recovery tools and are able to make changes.

## STEP-BY-STEP GUIDE TO START MACOS RECOVERY ON INTEL-BASED MACS

Hitting macOS Recovery on Intel-based Macs or older models sans Apple silicon opens up a plethora of important recovery tools devoted to reinstalling macOS, troubleshooting, and more. These Macs have three helpful combinations that each load a different version of macOS in Recovery mode. Here is a step-by-step process on how to start up from macOS Recovery with these models:

1) Power On Your Mac: Press the power button to turn on your Mac.
2) Press and Hold Correct Key Combination: Immediately after pressing the power button, press any of these key combinations and hold it (depending on the macOS version needed) to recover with
   a) Command ⌘ - R: These two keys, when held down, opens Recovery for the most recently installed macOS version on your Mac. It helps in the reinstallation of the current macOS without upgrading or downgrading.
   b) Option ⌥ - Command ⌘ - R: This combination attempts to load Recovery for the latest version of macOS compatible with your Mac. This can be used when you want to upgrade to the latest version and not erase any data.
   c) Shift (⇧) - Option (⌥) - Command (⌘) - R: The four keys hold down Recovery for the macOS version that originally came with your Mac, or the nearest version still available: This is the

best option if you want to install a fresh version of the original macOS installation.
3) Hold the buttons till Apple Logo / Spinning Globe shows:
   a) Continue to hold the keys till you see either the Apple logo or a spinning globe appear on-screen.
   b) Spinning globe: This usually points to an internet-based recovery. Such a recovery may occur when your Mac is unable to detect a local recovery partition on the disk.
4) Connect to Wi-Fi-if necessary:
   a) Of course, if you are prompted to select a Wi-Fi network, that simply means your Mac has to connect to the Internet so that it downloads recovery files.
   b) Wi-Fi menu: This is located on the top right corner of your screen, this is used in connecting your device to any Wi-Fi network. You can also connect your Mac to a network using an Ethernet cable for fast download speeds.
5) Log in using Admin Account if prompted:
   a) Sometimes, macOS will ask you which user account is an administrator and require you to enter the password.
   b) Select a correct user and click Next. It asks for an administrator password, actually the same password that you use to log in to the Mac.

That way, it ensures that Recovery tools cannot be accessed by any person who is not authorized or permitted to use them, besides giving the needed security to both your data and system integrity.

## REINSTALLATION OF MACOS FROM MACOS RECOVERY

1) **Start Up from Recovery**: Make sure you successfully start up from macOS Recovery. You will definitely be in the right place when you see the recovery window options such as "Reinstall macOS."

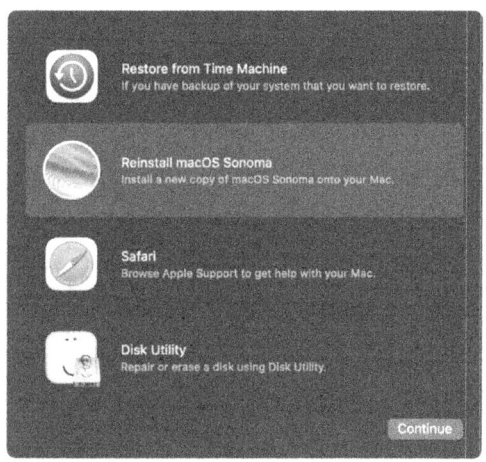

2) **Click the Reinstall macOS Option:** This usually sits quite boldly in the menu containing recovery options. Click it to begin the reinstallation process.
3) **Follow the On-Screen Instructions**
   a) Click Continue: After you have clicked on the reinstall option, click on the Continue option to proceed with the installation.
   b) Unlock Your Disk: After your Mac asks you to unlock your disk, it will ask for your password,

which will be required for logging into your Mac user account. It may also be the same password with which you logged in.

c) Choose Your Installation Destination: The installer will often give you a choice of disks. If your Mac has a volume marked 'Macintosh HD' and another called 'Macintosh HD – Data', then you should select 'Macintosh HD'. That's where making this selection will ensure the system's core operating system installs correctly. While the 'Data' volume houses your files and settings, the system files belong on the 'Macintosh HD'.

d) Disk Issues: This would be a problem when the operating system installer does not recognize your disk. You will have to erase the disk using pre-installer Disk Utility. Here's what to do: If Disk Utility is unable to erase the disk, it could point to a hardware problem. You are advised to consult with an Apple technician at this stage.

4) **Let Installation Complete:** Once the selection with respect to what is to be installed has been made, go ahead and let the installation process complete.

a) 'Do Not Interrupt': It is very important not to let the Mac sleep or shut the lid while installation is in process; interruption may cause partial installation and further create problems with the system.

b) Progress Indicators: Your MacBook will restart several times. You will see a progress bar in front of you on the screen showing how the installation is going. Note that there might be moments when nothing is shown on the screen, or the screen seems blank for many minutes- this is normal as the system works behind the scenes.

5) **Complete Setup Assistant**
   a) **Setup Assistant**: If you have not interrupted the installation with Time Machine, your Mac may boot to a setup assistant when it is done. This will walk you through some early setup tasks such as connecting to Wi-Fi and logging in with your Apple ID.
   b) **Selling/Giving Away**: To sell, trade in, or give your MacBook away :
      i) DO NOT go through the Setup Assistant - instead, Quit it.
      ii) Shut Down Your Mac: Once you have logged out, click Shut Down. Do this for a very important reason: it will allow the new owner of this Mac to start clean with their information during their setup process.

## IMPORTANT CONSIDERATIONS DURING REINSTALLATION

1) Backup beforehand: Although a macOS reinstall will not destroy your files and applications, it is a good habit to always back up your important data with Time Machine or some other backup solution

in advance, should something go wrong during the process.

2) Software Updates following Reinstallation: Once macOS has been successfully reinstated, go into the App Store or System Settings and perform any pending software updates. This will make sure the operating system is updated and that your applications run on the latest versions.

3) Internet Connection: During the process, the MacBook should not be disconnected from the net, especially when installing an early version of macOS, it will request to download some files.

4) Patience is Key: This whole process of installation might take some time depending on the specifications of your system and the size of the macOS version to be installed. Be patient, therefore, and let it go to the very end.

## TIPS FOR SUCCESSFUL MACOS RECOVERY

1) Internet Connection: In case Recovery needs to download something, then good internet is quite critical. Larger downloads tend to be faster using Ethernet instead of Wi-Fi.

2) Backup Data: Back up all of your most crucial files before you begin the process of reinstalling to eliminate any possibilities of data loss.

3) Power Source: Connect your Mac to a source of power in order not to experience an interruption of the process due to shut down.

4) Firmware: The Apple silicon Macs are with firmware having extra security. Recovery mode is thus completely locked up within your system for security reasons.

# INDEX

## A

Accessibility Categories 161
Accessibility Settings ..... 25, 161, 166, 178
Accessibility Tip .............. 52
Ai 23, 24, 117
Airpods ........................... 177
Ambient Light ............... 154
Anti-Glare Screen ............. 16
App Behavior ................. 140
App Menu ......................... 41
Apple Account ...... 100, 110, 114, 119, 121, 130
Apple ID ...... 27, 28, 72, 100, 101, 102, 104, 105, 108, 109, 110, 112, 114, 117, 118, 119, 130, 153, 201, 219
Apple Menu .... 40, 127, 148, 167, 169, 173, 175, 177, 178, 182, 183, 187, 191, 212
Apple Silicon .................. 213
Assign Apps to Spaces .. 138
Atypical Speech ..... 187, 188
Automatic Window Management ............. 191

Availability ................. 98, 176

## B

Backgrounds Sounds ... 179
Bluetooth ................... 77, 81

## C

Camera Features ............... 17
Capabilities ..................... 184
Closing an App ................ 55
Color Filters .. 162, 172, 173
Computer Account ......... 28
Configuration Options ....... 16
Connect ... 26, 159, 211, 216, 220
Connectivity ...................... 98
Connectivity Options ......... 19
Control Center ... 40, 69, 76, 77, 78, 79, 81, 82, 83, 151, 152, 202
Conversions ......... 87, 88, 89
Currency .................... 87, 88
Customizing the Dock .... 59

## D

Dark Mode ....... 35, 156, 157
Deleting Spaces ..... 144, 206
Design ............................. 184
Desktop ... 34, 35, 38, 39, 43, 58, 59, 61, 71, 72, 73, 74, 78, 79, 100, 108, 109, 110, 111, 112, 113, 123, 124, 125, 126, 127, 133, 134, 137, 139, 140, 141, 149, 154, 156, 159, 191, 192, 194, 198, 199, 200, 201, 202, 203, 205, 206, 207, 208, 210
Desktop Environment .. 126
Desktop Functionality .... 39
Desktop Spaces ............. 205
Desktop Theme ............... 34
Device Synchronization 111
Disk Encryption .............. 33
Display ............................. 17
Display Colors ............... 172
Dock ... 38, 42, 44, 54, 55, 56, 57, 58, 59, 60, 73, 79, 102, 126, 138, 141, 148, 156, 191, 193, 194, 204, 208, 209
Documents .... 38, 40, 41, 44, 45, 47, 48, 50, 54, 56, 84, 85, 87, 100, 101, 102, 108, 109, 110, 111, 112, 113, 127, 169, 172, 195, 197
Dynamic Desktop .. 154, 155

## E

Editing Files ..................... 44
External Display ... 158, 159

## F

Family Sharing ...... 107, 153
Family Sharing Settings
................................... 153
File ...................... 45, 47, 118
Files .... 52, 56, 110, 111, 112
FileVault ........................... 33
Find My App ................... 29
Find My Mac . 101, 121, 122
Finder . 42, 44, 45, 46, 47, 49, 50, 51, 52, 54, 57, 60, 105
Finder Sidebar ................ 46
Finder Window ............... 44
Focus ... 68, 69, 70, 142, 189, 207
Folder Views ................... 47
Full Screen .... 145, 167, 189, 190, 193, 194, 196, 209
Full-Screen Mode 189, 194, 196
Full-Screen View .......... 193

## G

Gallery View .............. 48, 50
Go Menu .......................... 52
Green Buttons ............... 207

223

## H

Hearing . 161, 163, 175, 177, 178
Hearing Accessibility Features ...................... 175
Hearing Assistance ....... 177
Help Menu ....................... 42
Hover Text ............. 171, 172

## I

iCloud .. 28, 34, 47, 100, 101, 102, 104, 105, 106, 107, 108, 109, 110, 111, 112, 113, 114, 115, 117, 118, 119, 120, 121, 122, 153
iCloud Content ...... 104, 105
iCloud Drive.... 47, 101, 104, 105, 108, 109, 110, 111, 112, 113
iCloud Features ..... 100, 104
iCloud Photos101, 114, 115, 118
iCloud Shared Photo Library ....... 114, 115, 117
iCloud Storage ....... 117, 153
iCloud Syncing .............. 120
iCloud+.. 105, 106, 107, 112, 117, 120
Intel-Based Macs .......... 215

## K

Key Settings................... 149
Keyboard .57, 62, 84, 93, 94, 135, 137, 164, 183, 184, 204, 206, 210, 212
Keyboard Shortcuts..... 135, 206, 210

## L

Live Captions 163, 175, 176
Live Speech ... 184, 185, 186
Location ....... 25, 29, 30, 155
Locking Your Screen .... 150
Login ........................ 27, 130

## M

M4 ...... 16, 20, 21, 22, 24, 213
M4 Max ............................ 21
M4 Pro .............................. 21
Mac Operating System Reinstallation ........... 211
Mac OS Interface ............ 38
MacOS Recovery.. 212, 213, 215, 217, 220
Managing Spaces .. 134, 146
Managing Windows .... 189, 203
Mask Ambient Noise.... 179
Measurement ............ 87, 89

Memoji .... 28, 127, 128, 129, 130
Menu Bar 38, 39, 81, 82, 92, 94, 126, 151, 193, 194
Microphone ......... 79, 80, 99
Migration Assistant ........ 27
Mission Control 57, 58, 131, 132, 134, 136, 141, 143, 144, 145, 147, 191, 192, 193, 198, 203, 204, 205, 206, 207
Mobility .................. 161, 164
Multiple Desktop Spaces ..................................... 192
Multi-Tasking ................ 190

# N

New Chips ........................... 17
Night Shift ............. 157, 158
Notification Center .. 61, 62, 63, 64, 65, 71, 72, 73, 74, 124, 125, 126, 200
Notification Preferences 64
Notification Settings . 64, 65
Notifications . 63, 64, 65, 66, 67, 149

# O

Open a File ....................... 54
Open an Application ....... 54
Organizing ...................... 45

# P

Pair Hearing Devices .. 176, 177
Password ....................... 150
PC ..................... 27, 110, 118
Performance ...................... 21
Photos ..... 22, 101, 104, 105, 114, 115, 116, 118, 123, 125
Pointer .. 156, 162, 165, 172, 182, 183
Preview 49, 52, 82, 129, 189, 208
Preview Pane ................... 49
Price ................................. 16
Privacy and Security ..... 118
Purchases ...................... 119

# Q

Quick Actions ...... 50, 52, 86
Quick Look ................ 51, 52
Quick View ...................... 44

# R

Recording Indicator ....... 79
Red ......................... 173, 207
Reinstallation 211, 217, 219, 220

225

## S

Screen Reader ............... 168
Screen Saver .. 133, 149, 151
Screen Time ............. 31, 153
Security ...... 80, 99, 122, 155
Setting Up 25, 66, 68, 91, 93, 100, 104, 114, 115, 121, 154, 157
Setting Up iCloud. 100, 104, 114
Sharing Photos ...... 114, 116
Signing In ....................... 100
Siri 23, 29, 31, 32, 33, 87, 91, 92, 93, 94, 95, 96, 97, 98, 99, 169, 187, 188
Siri Commands ................ 97
Siri Suggestions ............... 91
Software ... 36, 150, 152, 220
Spaces .... 131, 132, 133, 134, 136, 137, 138, 140, 141, 142, 143, 144, 145, 146, 147, 192, 193, 204, 205, 206
Speech ... 161, 165, 185, 186, 188
Speed Boost ....................... 16
Split View ...... 144, 145, 147, 190, 194, 195, 196, 199, 205, 209, 210
Spotlight .. 39, 40, 55, 73, 84, 85, 86, 87, 88, 89, 90, 91, 92, 93, 95, 96

Stage Manager . 78, 79, 191, 201, 202, 203
Storage .. 105, 107, 112, 115, 120, 153
Storing Files ................... 111
Syncing ..................... 45, 108
Syncing Devices .............. 45
System Settings .. 36, 39, 59, 60, 64, 65, 68, 79, 80, 81, 87, 89, 91, 92, 93, 95, 96, 99, 100, 104, 105, 106, 110, 112, 114, 119, 121, 123, 124, 126, 127, 133, 135, 137, 138, 141, 148, 150, 151, 152, 153, 154, 155, 156, 158, 159, 161, 167, 169, 171, 173, 175, 177, 178, 179, 181, 182, 183, 185, 186, 187, 191, 194, 220

## T

Technology Advancements 21
Temperature 87, 88, 89, 158
Third-Party Widgets ...... 73
Tiling Windows ............. 196
Time Machine 27, 211, 212, 219
Time-Based Visual Changes ..................... 154
Touch ID ... 34, 93, 174, 212, 213

Transfer Data ...................27
Troubleshooting Siri.......98
True Tone.......................154
Turnoff Your Mac .........211

## U

Up To Date.....................150
Update Your Mac............35
Updating macOS...........152
Using the Dock................59

## V

Virtual Desktops ...........192
Vision.......................161, 162
Voice Control. 164, 181, 182
Voice Shortcuts .............186
Voiceover .......................168
VoiceOver Features ......169

## W

Wallpaper 39, 123, 124, 133, 134, 149, 151, 154, 155, 198
Warmer Colors.............. 157
Widget 62, 65, 71, 72, 73, 74, 124, 125, 126, 199, 200, 201
Wi-Fi 26, 40, 77, 81, 98, 102, 216, 219, 220
Window Management Tips ...................................... 198

## Y

Yellow............. 173, 207, 208

## Z

Zoom ..... 142, 162, 166, 167, 168, 171, 172
Zoom In ......................... 171

227

Made in the USA
Las Vegas, NV
05 March 2025